A NEW SCHOOL GEOGRAPHY
Volume 5

NEW SCHOOL SERIES

GENERAL EDITOR (Science): R. Stone, M.A., A.Inst.P.
Second Master, Manchester Grammar School

GENERAL EDITOR (Arts): B. A. Phythian, M.A., B.Litt.
Headmaster, Langley Park School for Boys, Beckenham

F. R. Dobson, B.A.
H. E. Virgo, M.A.
THE ELEMENTS OF GEOGRAPHY IN COLOUR

A New School Geography

Volume 1 THE ELEMENTS OF GEOGRAPHY
Volume 2 THE BRITISH ISLES
Volume 3 MAP READING AND LOCAL STUDIES
Volume 4 NORTH WEST EUROPE
Volume 5 CANADA AND THE UNITED STATES

A New Geology

M. Bradshaw, M.A.

A New Chemistry

S. Clynes, B.Sc., F.R.I.C.
D. J. W. Williams, M.A.
J. S. Clarke, B.Sc., M.A., F.R.I.C.

A New Biology

K. G. Brocklehurst, M.A., M.I.Biol.
Helen Ward, B.Sc.

General School Physics

R. Stone, M.A., A.Inst.P.
N. Bronner, M.A.
Volume 1 HEAT AND LIGHT
Volume 2 MECHANICS, HYDROSTATICS AND SOUND
Volume 3 MAGNETISM AND ELECTRICITY

Le Français d'Aujourd'Hui

Parts One, Two, Three, Four (CSE) and Four (GCE)
PUPIL'S BOOK, TEACHER'S HANDBOOK, AUDIO VISUAL COURSE
P. J. Downes, M.A.
E. A. Griffith, B.A.
P. Houldsworth, Ph.D. *(CSE volume only)*

Starting Points

A New English Book
G. P. Fox, M.A.
B. A. Phythian, M.A., B.Litt.

Listen

An Anthology of Dramatic Monologues
A. Thompson, M.A.

Storylines

A Teaching Anthology of Short Stories
A. Thompson, M.A.

Touchstones

A Teaching Anthology of Poetry (In Five Volumes)
M. G. Benton, M.A.
P. Benton, M.A.

Creeds and Controversies

A General Studies Course on the Nature of Belief
Rev. P. F. Miller
Rev. K. S. Pound

A NEW SCHOOL GEOGRAPHY Volume 5

CANADA AND THE UNITED STATES

F. R. DOBSON, B.A.

Formerly Senior Geography Master
The Forest Grammar School
Winnersh, Berkshire

H. E. VIRGO, M.A.

Deputy Headmaster
The Grammar School
Hinckley, Leicestershire

THE ENGLISH UNIVERSITIES PRESS LIMITED

ISBN 0 340 05060 8

First published 1970
Reprinted 1972, 1974

The English Universities Press Ltd
St Paul's House, Warwick Lane, London EC4P 4AH

Printed in Great Britain by Butler & Tanner Ltd,
Frome and London

CONTENTS

EDITOR'S INTRODUCTION

We live in a rapidly changing world: some might say in a too rapidly changing one. Today ideas can alter as much in a decade as in the whole of the preceding century. As the kaleidoscopic patterns of syllabus and method pass before us, it is not always easy to distinguish the important and permanent from the ephemeral. It is our hope that, in the New School Series, we shall be able to produce a group of books which will help colleagues in the classroom and their pupils to meet the challenge of the next ten years.

Our main purpose is to provide a wide range of books, covering both the Arts and the Sciences at a number of levels, which will allow the teacher increased latitude in his approach and offer him full scope to develop the creative aspects of his work. It must not be forgotten that the suitability of a particular text for a given form depends largely on the way in which the teacher uses it: the speed and depth of the work will be dictated by the abilities and interests of the pupils rather than by the wishes of the teacher or the nature of the text. Many of the new books will be suitable for all but the lower streams of the comprehensive schools: others will express the newer conceptual, as opposed to factual, approach to teaching but may be contained within somewhat narrower academic boundaries. We shall be greatly in debt to the many teachers, professional bodies, and others whose untiring efforts have done so much to change the pattern of teaching in recent years. Nor must we forget those pupils who, whether consciously or not, have been the guinea-pigs in the experiments which were necessary to prove the new ideas.

We hope to arrange for books to be written by teams of two or more experienced teachers who have tried out the new methods and syllabuses in the classroom, and who will be able to engender in their readers the same enthusiasm which they have instilled into their pupils. It is this dual interest of those who teach and those who are taught which is the key to all successful learning, a process in which we hope to play our part.

R. STONE

PREFACE

This book is designed for use in the fourth or fifth forms of secondary schools. As in previous volumes, we have aimed to present the subject matter in a concise and systematic form in order to assist the pupil in his task of learning and understanding. We have devoted much space to maps, diagrams and photographs, which are intended to play a vital part both in illustrating the text and in supplying additional information. All place names mentioned in the text appear on the maps, so that reference to an atlas should be unnecessary. Many of the photographs are accompanied by questions which, it is hoped, will stimulate classroom discussion. In addition, there are written exercises at the end of each chapter, graded to cater for a variety of abilities.

Every effort has been made to provide up-to-date information, particularly on recent economic developments including the increasing part played by mineral oil, natural gas and hydro-electricity in the power supplies of Canada and the United States. The growth of new manufacturing industries in Canada, the growing dependence of the United States on imported raw materials, and the close economic co-operation between the two countries, are other important topics.

We should like to express our appreciation to the officials of Canada House and the United States Information Office in London for help in supplying information and for the loan of photographs. Our thanks are also due to our editor, Mr R. Stone, for his useful suggestions.

F.R.D.
H.E.V.

1: INTRODUCTION – PHYSICAL GEOGRAPHY

The Building of the Land

North America's physical features are of great variety and range from the rugged Western Cordillera, which rise well above the snow line, and the more subdued Appalachian Mountains in the east, to the ancient, worn-down Canadian Shield in the north-east, and the more recently formed Interior Plains which stretch from the Arctic Ocean to the Gulf of Mexico.

The geological history of North America had much in common with that of other continents, and consisted of long, quiet periods during which sedimentary rocks, thousands of metres thick, accumulated mostly on the sea bed or in lakes. These sedimentary rocks were then subjected to enormous pressures which forced them up into folds, the folding movements being accompanied by widespread igneous activity. It was in this way that the great mountain ranges came into being, and there is evidence of at least three periods of mountain building, roughly corresponding to the Caledonian, Hercynian and Alpine movements in Europe.

Early fold mountains covered the region now occupied by the Canadian Shield, but these have long since been worn away. Of later origin were the Appalachians, first raised up about the time of the Hercynian mountain building period in Europe, and the Western Cordillera which were formed in the most recent period of uplift, the Alpine.

The Ice Age

As in northern Europe much of North America was covered by an ice sheet during the Ice Age. The limits of this sheet fluctuated considerably and its maximum extent is shown in Figure 1.2. During this period vast amounts of material were carried across the lowlands, much of it to be deposited as boulder clay, i.e. a mixture of stiff clay interspersed with boulders of all sizes. An extensive series of moraines mark the southern limit of the ice at various stages in its retreat, and as the ice melted enormous volumes of water were released. Some of this accumulated between the edge of the ice sheet and the surrounding higher land to form huge lakes.

The Great Lakes (Lakes Ontario, Erie, Huron, Superior and Michigan) were formed in this way, although the hollows they occupy are pre-glacial in origin. During the period of maximum glaciation the melt water escaped into the Mississippi drainage system, but as the ice withdrew northwards, it found new outlets, including the Mohawk and Hudson valleys, and later the St Lawrence (Figure 1.1). With the final retreat of the ice, the lakes took on their present form.

Today there is little permanent ice except on the islands to the north of the Canadian mainland and in the glaciers of Alaska and other parts of the Western Cordillera, which represent only relics of the massive glaciers which produced such impressive erosive features in these mountains in the past.

Recent fluctuations in the levels of land and sea have had a marked effect on North America's coastline. Following the melting of the ice, the rise in sea level led to the drowning of the outer ranges along the north Pacific coast to form numerous islands, and the submergence of glaciated valleys to produce fiords. The Atlantic coast also shows much evidence of drowning, but to the south of New York an opposite movement has taken place and a broad coastal plain has been formed by the emergence of the continental shelf.

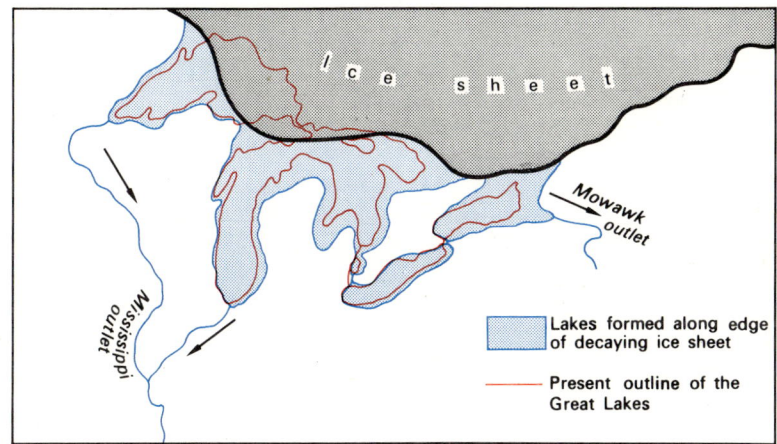

FIG. 1.1 **A stage in the formation of the Great Lakes towards the end of the Ice Age**

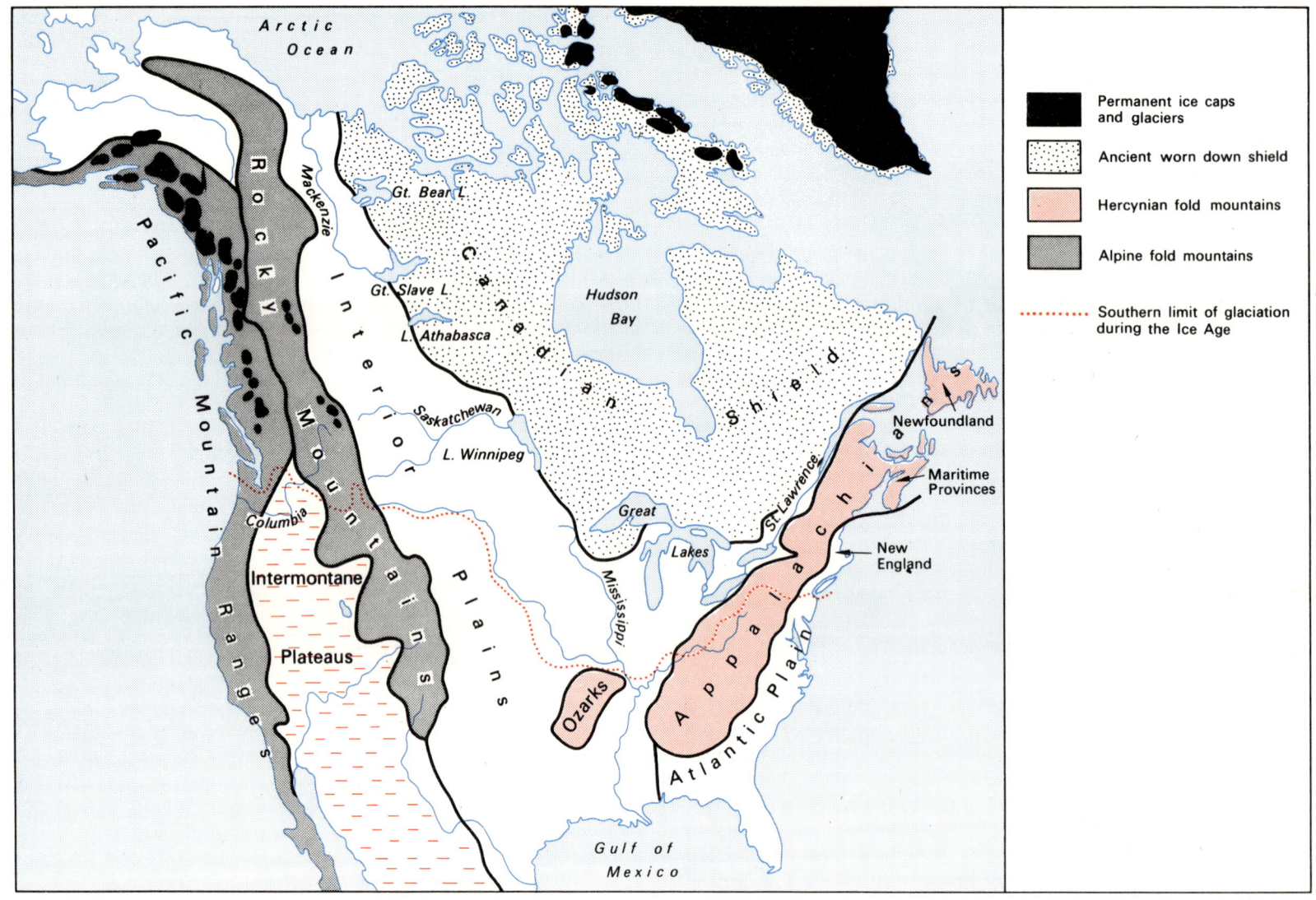

FIG. I.2 **Physical features of Canada and the United States**

Structural Regions

The oldest part of North America, the Canadian Shield, lies in the north-east and extends from the Hudson Bay region as far as the St Lawrence, Great Lakes, Lakes Winnipeg and Athabasca, and the Great Bear and Great Slave Lakes. The Shield is a low plateau (peneplain) which has been worn down by rain, rivers and other agents of erosion over millions of years. Glaciation has produced innumerable hollows so that the surface is covered with a maze of lakes and water-courses. The rock, which is ancient and crystalline, is pierced in places by valuable mineral veins, and throughout much of geological history it has acted as a resistant block round whose margin softer sediments have been crumpled into fold mountain ranges.

To the south-east of the Canadian Shield lie the Appalachian Mountains. These ancient fold mountains were worn down almost to a peneplain and then uplifted by earth movements. The general trend of the mountains is from south-west to north-east, roughly parallel with the Atlantic coast. The oldest rocks occur in New-foundland, the Maritime Provinces of Canada, northern New England, and the Piedmont zone south of New York, with a more recently folded zone to the west. The Appalachians rise to almost 2000 metres above sea level and form an important barrier to com-munications except for the easy passes following the Hudson and Mohawk valleys.

The Interior Plains form a continuous belt from the Arctic Ocean to the Gulf of Mexico, and are drained by a number of great rivers including the Mackenzie, Saskatchewan and Mississippi. The plains are built of almost horizontally bedded strata, younger in age than the rocks which form the Appalachians except for a few 'islands' of older rocks which protrude from the plain, the chief of which are the Ozark and Ouachita hills of the south-central United States. The youngest rocks of all are in the south where they form the broad lowlands bordering the Gulf of Mexico and the southern Atlantic coast. To the west of the Mississippi the land rises gradually to reach about 1500 metres in the High Plains bordering the foothills of the Rocky Mountains.

The Western Cordillera occupy one third of North America and are a complex system of young Alpine fold mountains with a general trend from north to south almost parallel to the Pacific coast. In the east are the Rocky Mountains which form some of the most impressive scenery in the continent and in the west are the Pacific mountain ranges which include the Coast Ranges, Sierra Nevada and Cascades.

Between these eastern and western systems are a series of inter-montane plateaus and basins which have resulted from elevation and subsidence on a massive scale often associated with faulting. Wide-spread volcanic activity has left numerous extinct volcanoes and even a few still-active ones in Alaska, including Mt McKinley (6192 metres). There were also outpourings of basalt through large fissures, forming extensive plateaus, the largest being that of the Columbia-Snake basin.

The Climate and Vegetation of Canada and the United States

The following major factors have a great influence on the climate of Canada and the United States:

(1) *Differences of Latitude*. The amount of heat received from the sun at any place depends on the angle of the sun's rays and the length of daylight, both of which vary with latitude. Canada and the United States cover a wide range of latitude, extending from the Arctic ice cap to the sub-tropical zone.

(2) *Influence of the Sea*. The sea takes longer than the land to heat up in summer and cool down in winter. It has therefore a moderating effect on the temperatures of coastal areas, which are said to have a maritime climate, in contrast to interior areas which have a continental climate with extremes of temperature. Ocean currents also have a marked effect. North of latitude 40° north the west coast is influenced by the warm North Pacific Current and the east coast by the cold Labrador Current. South of latitude 40° north the cool Californian Current influences the west coast, and the warm Gulf Stream the east.

(3) *Winds and Pressure*. Winds have a considerable effect on tem-perature and also carry moisture from the sea, resulting in precipita-

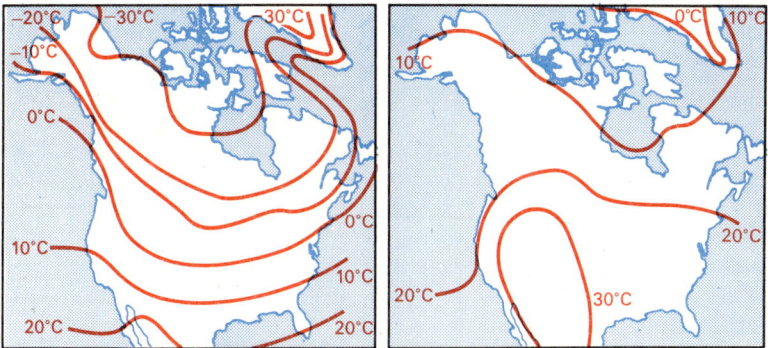

FIG. 1.3 **Isotherms for January and July**

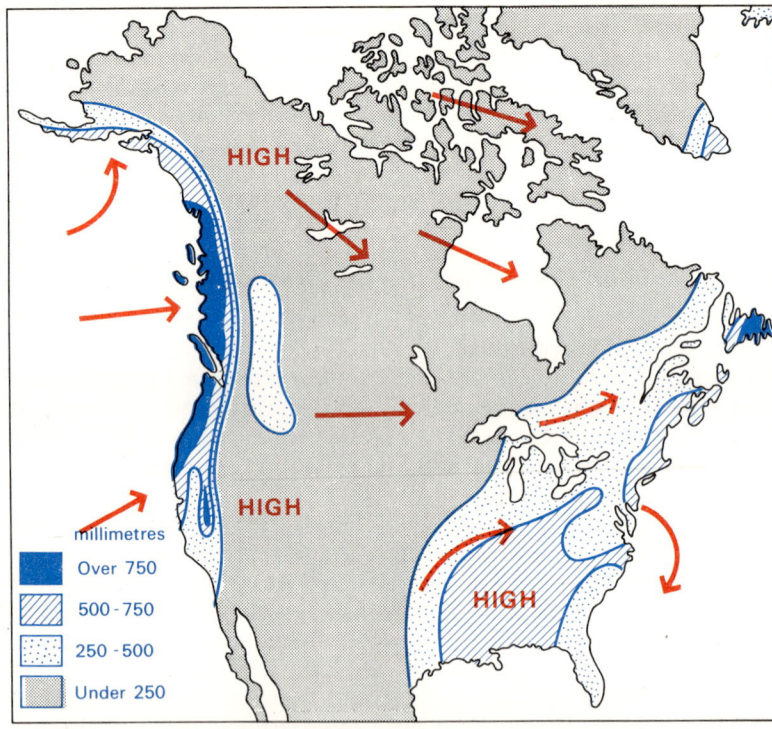

FIG. I.4A **Prevailing winds, pressure systems and rainfall from November 1st to April 30th**

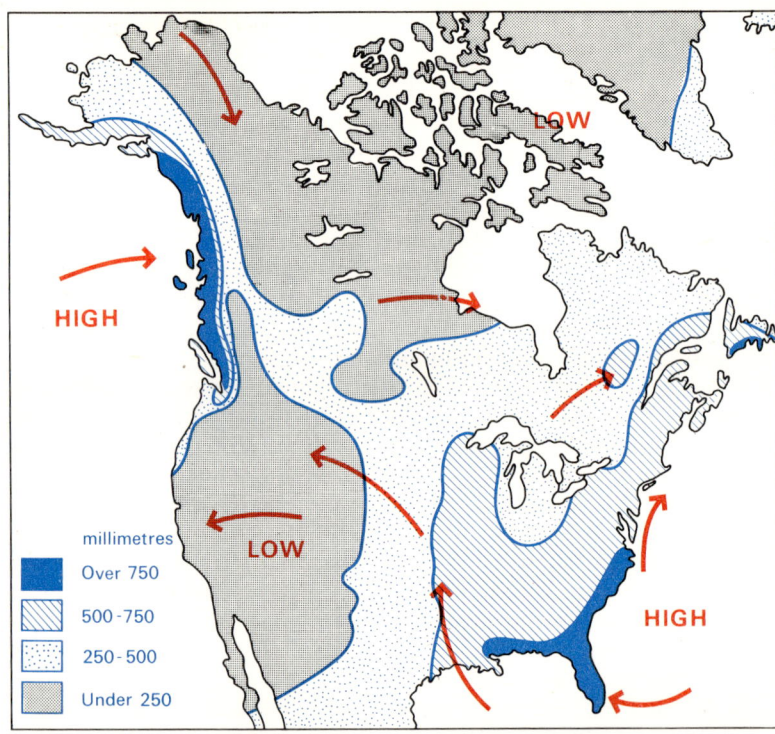

FIG. I.4B **Prevailing winds, pressure systems and rainfall from May 1st to October 31st**

tion over the land. North America comes under the influence of the following world wind belts: (a) South-Westerlies all the year round to the north of latitude 40° north, (b) North-East Trades all the year round to the south of latitude 30° north, and (c) South-Westerlies in winter and North-East Trades in summer between latitudes 30° and 40° north.

However, the actual wind directions are not as simple as this, and depend largely on seasonal variations of pressure over the land mass. In winter the continental interior is dominated by a high pressure system which causes outblowing winds, whilst in summer this is replaced by a low pressure system, with inblowing winds.

(4) *Relief.* Temperatures are lower and precipitation greater over high ground. In addition, the north-south trend of the western and eastern mountain ranges tends to exclude maritime influences from

the interior. This is particularly so in the west, where the Western Cordillera acts as a barrier against the mild, moist westerly winds. On the other hand, the Central Plains are open to the north and south, and are thus exposed to strong air movements which bring heat waves from the south in summer and cold waves from the Arctic in winter.

Climate and Vegetation Regions

1. *Sub-Tropical Regions.* Small portions of the south-western and south-eastern United States have sub-tropical conditions.

(a) In the south-west, in Arizona and the extreme south of California, is a hot desert region. Since it lies on the leeward side of the continent in the Trade Wind belt, the climate is almost rainless. Summers are very hot and winters warm, and the lack of cloud allows

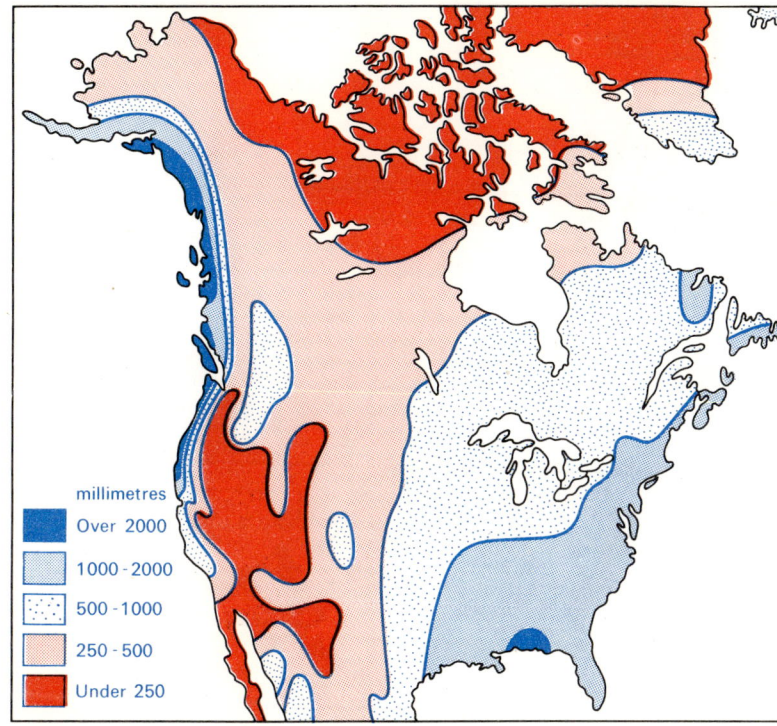

FIG. I.5. **Annual rainfall**

millimetres
- Over 2000
- 1000 – 2000
- 500 – 1000
- 250 – 500
- Under 250

rapid radiation, so that hot days are followed by cool nights. The natural vegetation is scanty, with marked adaptations to counter the drought, and includes a variety of cacti and small shrubs such as the creosote bush.

(b) In the south-east, in distinct contrast, are the coastal fringes of the Gulf of Mexico and the peninsula of Florida, which have a hot, equable climate with considerable precipitation resulting from on-shore winds. Vegetation is luxuriant and includes the southern pine, cypress and mangrove swamps.

2. *Warm Temperate Regions*. These regions lie between latitudes 30 and 40° north, and consist of:

(a) A Warm Temperate Western Margin region in California, with a 'Mediterranean' type of climate. Winters are mild and rainy, due to

onshore South-Westerly winds, and summers are hot and dry, with offshore Trade Winds. The natural vegetation is adapted to withstand the summer drought and consists of deep-rooted, coarse grass and poor quality woodland known as chaparral, but there is tall coniferous forest on some of the mountain slopes.

(b) A Warm Temperate Eastern Margin region, extending from the Atlantic coast to the foothills of the Rockies. This region has similar temperatures to the western margin, but has heavier and more evenly-distributed rainfall. In winter the rainfall is caused by cyclonic disturbances, while in summer it is brought by moist winds from the sea which blow towards the low pressure over the interior of the continent. The abundant rainfall has produced a natural vegetation of thick woodland, with both evergreen and deciduous trees including oak, beech and hickory.

3. *Cool Temperate Regions*

(a) A Cool Temperate Western Margin, which extends from latitude 40 to 60° north in Oregon, Washington and British Columbia. The region is under the influence of the warm North Pacific Drift and moist South-Westerly winds, with associated depressions. It has a 'West European' type of climate, characterised by mild winters, cool summers, and heavy, well-distributed rainfall, but high mountains running parallel to the coast prevent the maritime influences from penetrating far inland. The natural vegetation is mainly coniferous forest including the tall, straight Douglas Fir.

(b) A Cool Temperate Eastern Margin, between latitudes 40 and 50° north, covering the Great Lakes, St Lawrence Valley and New England. In winter mean temperatures are below freezing point and cold westerly winds blow from the continental interior, whilst the chilling effect of the Labrador Current causes the sea to freeze as far south as Newfoundland. In summer temperatures are slightly higher than in the western margin and the winds are more variable, sometimes blowing from the Atlantic towards the low pressure of the interior. Precipitation is well distributed and is mainly caused by the frequent depressions. The natural vegetation consists of coniferous forest in the north and deciduous trees including maple in the southern parts.

4. *Temperate Interior Region*

The interior of the continent is remote from the moderating influence of the oceans so that great extremes of temperature are experienced. Winters are very cold, with mean temperatures well below freezing

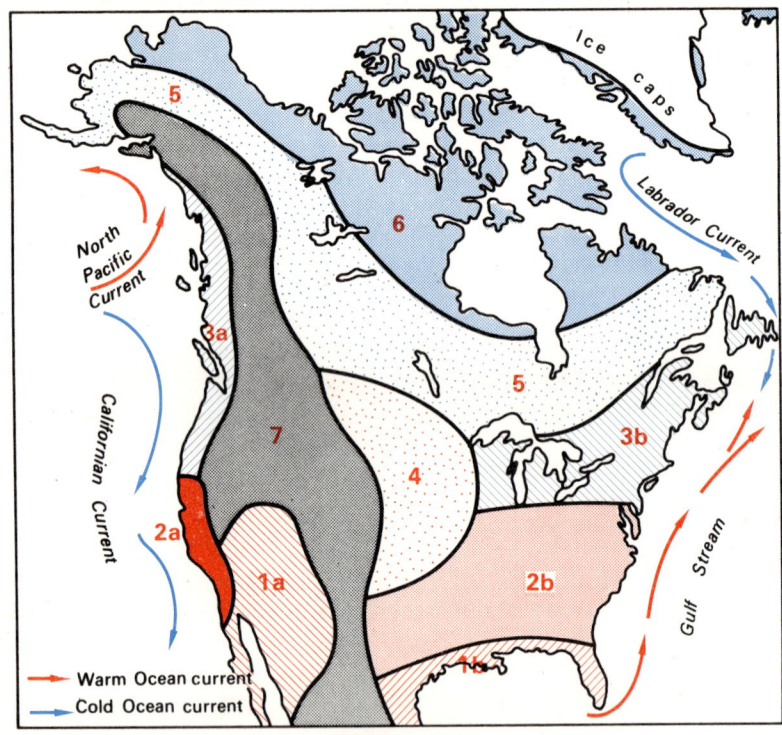

FIG. 1.6 **Climatic Regions**

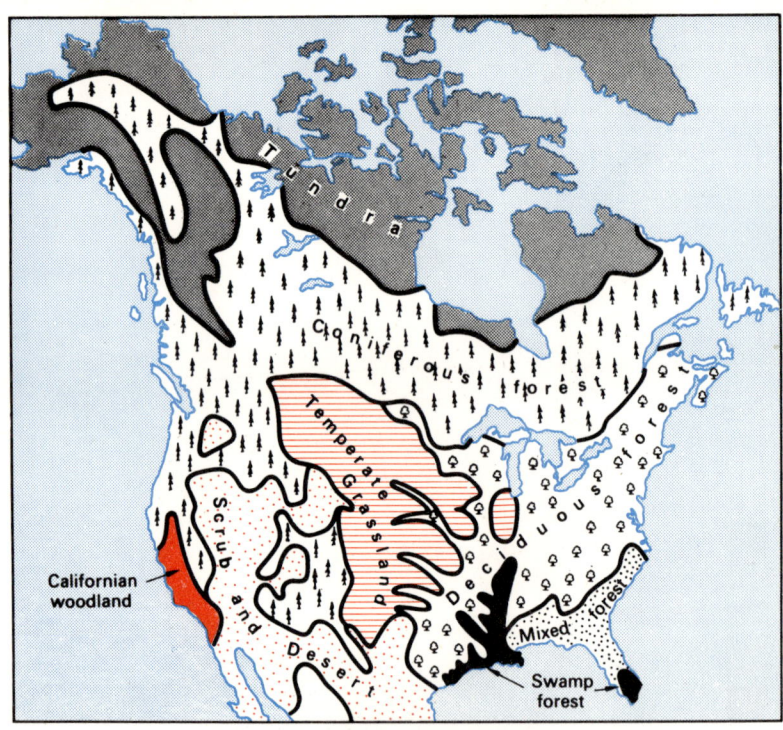

FIG. 1.7 **Natural Vegetation**

point, and high pressure causes outblowing winds so that there is very little precipitation. Summers are warm and pressure is low, causing inblowing winds and light rainfall, much of it in the form of thunderstorms. The light rainfall and high rate of evaporation in summer have led to a natural vegetation of quick-growing grass which withers in late summer (hence the name 'Prairies'). Few trees can tolerate these conditions.

5. *Cold Temperate Region*
This region stretches across the continent from Labrador to Alaska and is characterised by long, severe winters and short, mild summers. Precipitation consists of light falls of snow in winter and occasional showers of rain in summer. Since evaporation is small, there is sufficient soil moisture for coniferous forest, including spruce, fir, pine

and larch, which are able to survive under these conditions.

6. *Tundra*
Farther north still is the tundra, a region of long, bitterly cold winters with very little daylight during November, December and January, and short, cool summers. Precipitation is light and is mainly in the form of snow. Although the topsoil thaws in summer, the subsoil remains frozen, a condition known as permafrost. The natural vegetation is restricted to mosses, lichens, a variety of small shrubs including cranberry and bilberry, and quick-growing flowers and grasses which lie dormant for most of the year.

7. The mountains, and the Western Cordillera in particular, have an alpine climate with marked variations in temperature, precipitation and natural vegetation, depending on altitude.

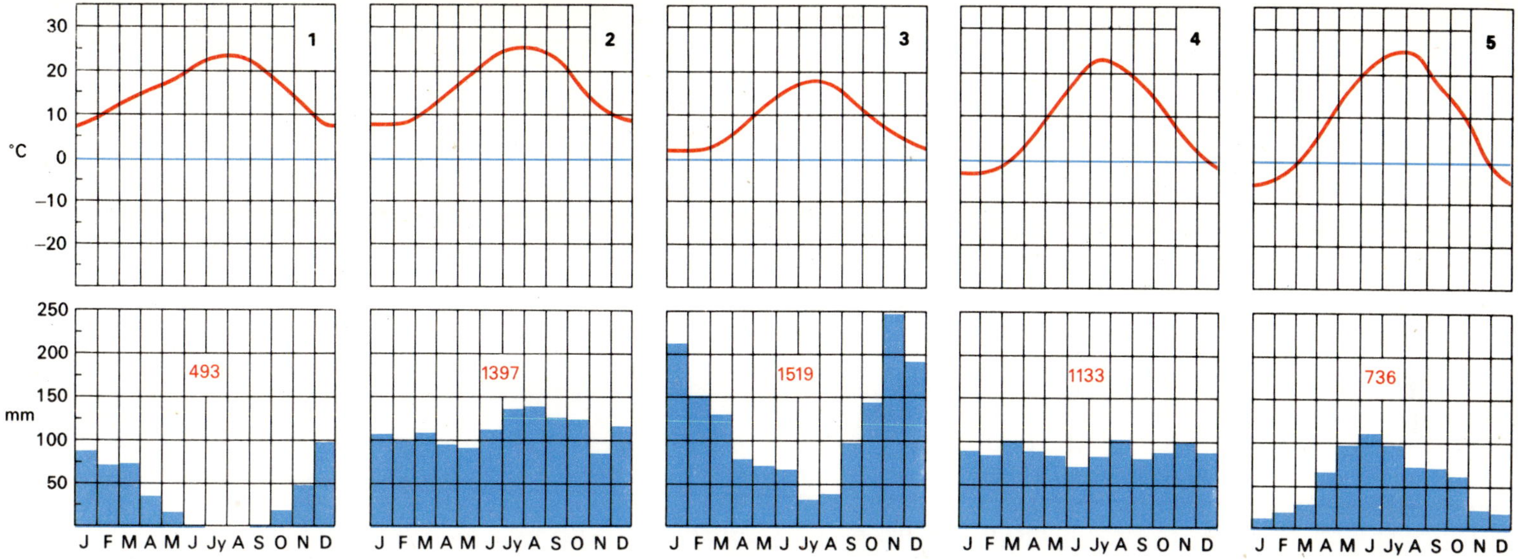

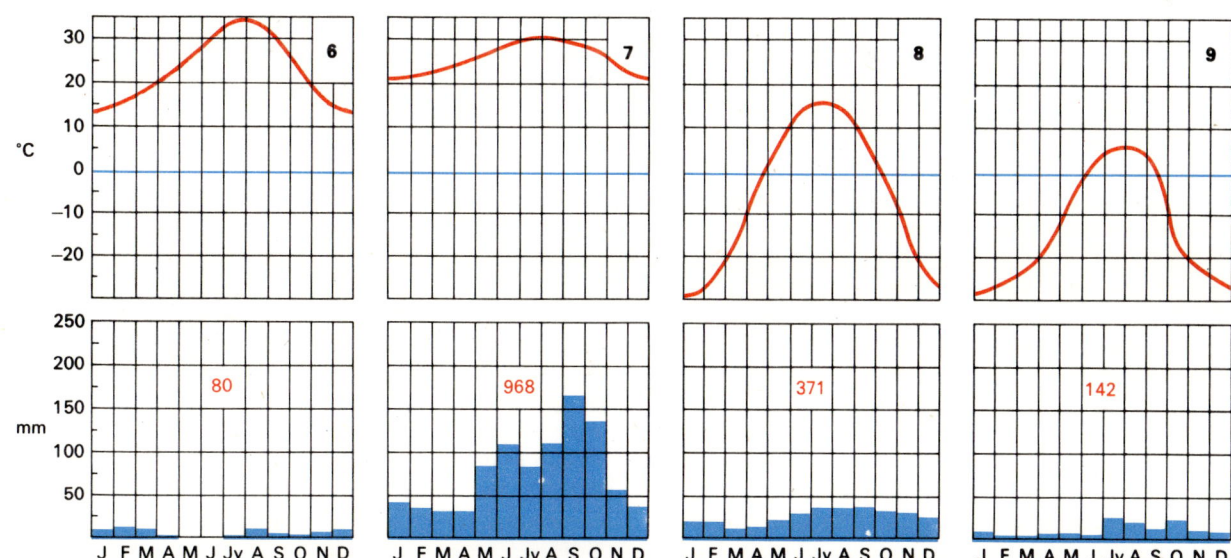

FIG. 1.8 **Climatic graphs**

These graphs represent the nine distinctive climatic types outlined above. Describe the climate for each station and then suggest which region it is in. The stations are Vancouver, Omaha, Yuma (Arizona), Key West, Hatteras, Barrow Point (Alaska), Dawson, Sacramento and Boston, but not in that order.

2: CANADA – GENERAL ECONOMIC SURVEY

The opening up of Canada can be said to date from the voyages of the Frenchman Jacques Cartier who, in 1535, sailed up the St Lawrence as far as the present site of Montreal. French traders followed, and many of them established friendly relations with Indian tribes for the barter of furs. During the seventeenth and eighteenth centuries French farmers arrived in large numbers, settling mainly in the St Lawrence Lowland, and about the same time small, scattered fishing communities were making their appearance on some of the islands near the mouth of the St Lawrence.

Meanwhile, British settlers occupied parts of Newfoundland, the Maritime Provinces, the Hudson Bay region and southern Ontario. There was considerable hostility between the British and French, leading to bitter fighting during the Seven Years' War, which culminated in Wolfe's capture of Quebec in 1759. The Peace of Paris was signed in 1763 and confirmed Britain's possession of all Canada except for the small islands of St Pierre and Miquelon, but a large French population remained in Quebec.

Canada today has a federal system of government and consists of ten provinces – Newfoundland, the Maritime Provinces (Prince Edward Island, Nova Scotia and New Brunswick), Quebec, Ontario, the Prairie Provinces (Manitoba, Saskatchewan and Alberta), and British Columbia. Each province has its own government with certain powers over taxation, health, welfare, education, etc., but all-Canada affairs come under the federal parliament which meets in Ottawa. The Yukon and North-West Territories, because of their small populations, are not provinces and are administered by the federal government.

With an area of 9 220 000 square kilometres, Canada is the third largest country in the world, after the U.S.S.R. and China, but its population numbers only about 21 million, giving an average density of about 2¼ to the square kilometre. The greater part of the country is too cold, mountainous or barren to attract settlement, and the population is concentrated close to the United States border, nearly three-quarters in eastern Canada, a fifth in the Prairie Provinces, and a tenth in British Columbia. Of the total population, 45 per cent are of British stock and 30 per cent French (most of whom live in the St Lawrence valley). The rest came mainly from other European countries but also include some 220 000 Indians and 15 000 Eskimos.

Canada's Resources

Canada is very well endowed with natural resources, and as a result of the efficient use of these resources, the people have a high standard of living. Nearly half the country is forested, mainly with coniferous trees, and there is abundant and varied mineral wealth. The coastal waters, rivers and lakes are rich in fish, and the rivers and lakes also provide great reserves of hydro-electric power. The amount of land suitable for farming is small when compared with the total area of the country, but is nevertheless sufficient to feed the whole population and to produce a large surplus of food for export.

The Forests

Pioneers were first attracted to the forests for the fur-bearing animals rather than for the timber, and trading companies, which included the Hudson's Bay Company, established trading posts to purchase the pelts of squirrel, fox, ermine, mink, marten, beaver, muskrat, bear and other animals from Indian, Eskimo and European trappers. Today the number of fur-bearing animals has dwindled, and large quantities of furs are now produced by breeding the animals in captivity.

Canada has the world's second largest reserves of softwood timber (after the U.S.S.R.), the commonest trees being spruce, fir, pine, hemlock and cedar. Prior to 1930 most of the timber was cut in the eastern provinces, but since then the west coast has greatly expanded its production. British Columbia, which has the finest constructional timber including Douglas Fir, now contributes two-thirds of the country's output. Quebec, Ontario and Newfoundland still provide most of the pulp and paper, spruce being the chief source for this. Among the valuable by-products are cellulose, rayon, resin and turpentine.

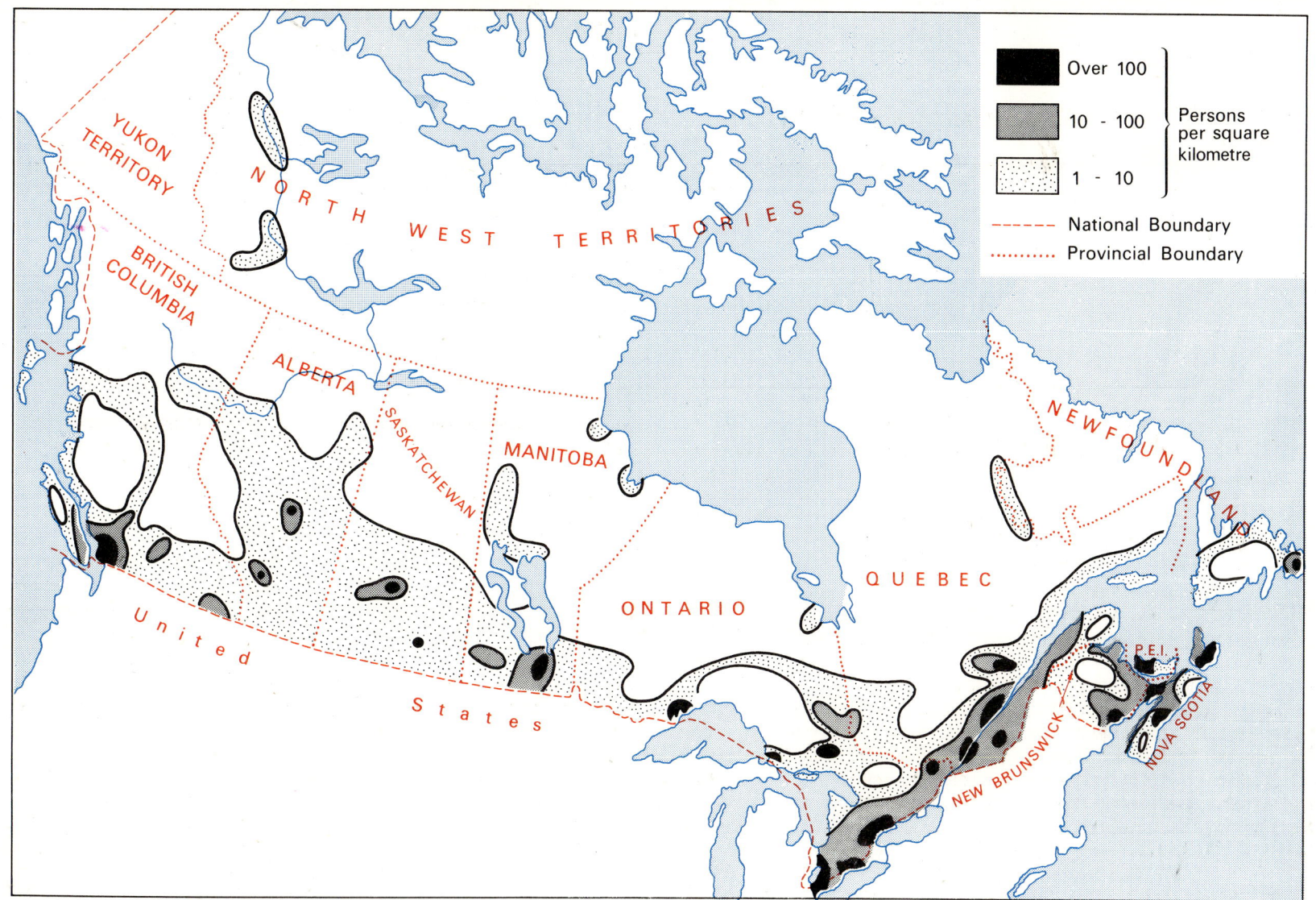

FIG. 2.1 **Canada – distribution of population**

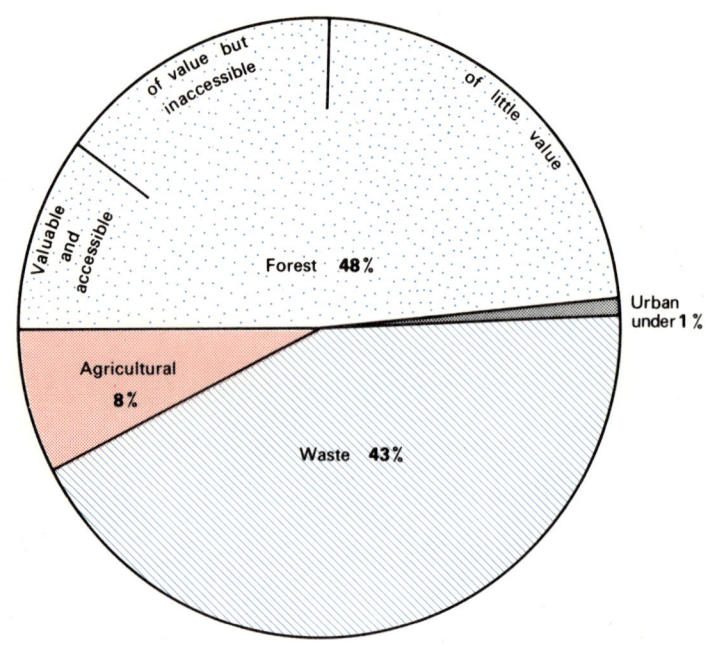

FIG. 2.2 **Canada – land use**

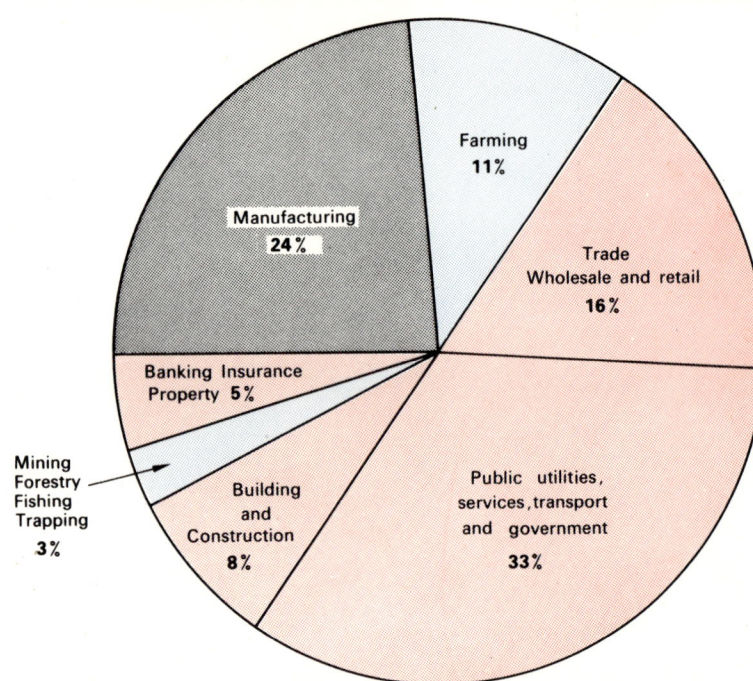

FIG. 2.3 **Canada – employment (percentage of working population)**

Fishing

Fishing is an important occupation off both the east and west coasts of Canada, as well as in the lakes and rivers, and about two-thirds of the catch is exported. Off the east coast the broad continental shelf, which includes the Grand Banks, is an ideal breeding ground for fish, and the meeting of the Gulf Stream and Labrador Current encourages the growth of plankton, the chief food of fish. The numerous coastal inlets are excellent natural harbours and the chief fish caught are cod and herring, with lobsters in inshore waters. Off the west coast the chief catch is salmon, most of which are netted off the river mouths, followed by halibut.

Farming

The short growing season and infertile soils are serious limiting factors for agriculture over much of Canada, but conditions are more favourable in the south. There is intensive mixed farming, with an emphasis on dairying, in the Ontario Peninsula, in the St Lawrence Lowland, and in the smaller lowlands and valleys of the Maritime Provinces and British Columbia. In the Prairie Provinces extensive wheat farming is carried on, with large-scale cattle and sheep ranching in the drier parts. Deciduous fruits are of particular importance in the Annapolis Valley of Nova Scotia, along the lake shores of south-east Ontario, and in the southern valleys of British Columbia.

Almost all the farms are privately owned and they vary considerably in size, being smallest among the French population of the St Lawrence Valley, and largest in the Prairies. Since the war farms have become larger and fewer in number, and output has grown in spite of a smaller number of farm workers. This has been made possible by increased mechanisation and electrification, and the greater use of fertilisers and other chemical aids. Methods are becoming more intensive even in the Prairies, where mixed farming is tending to replace arable farming.

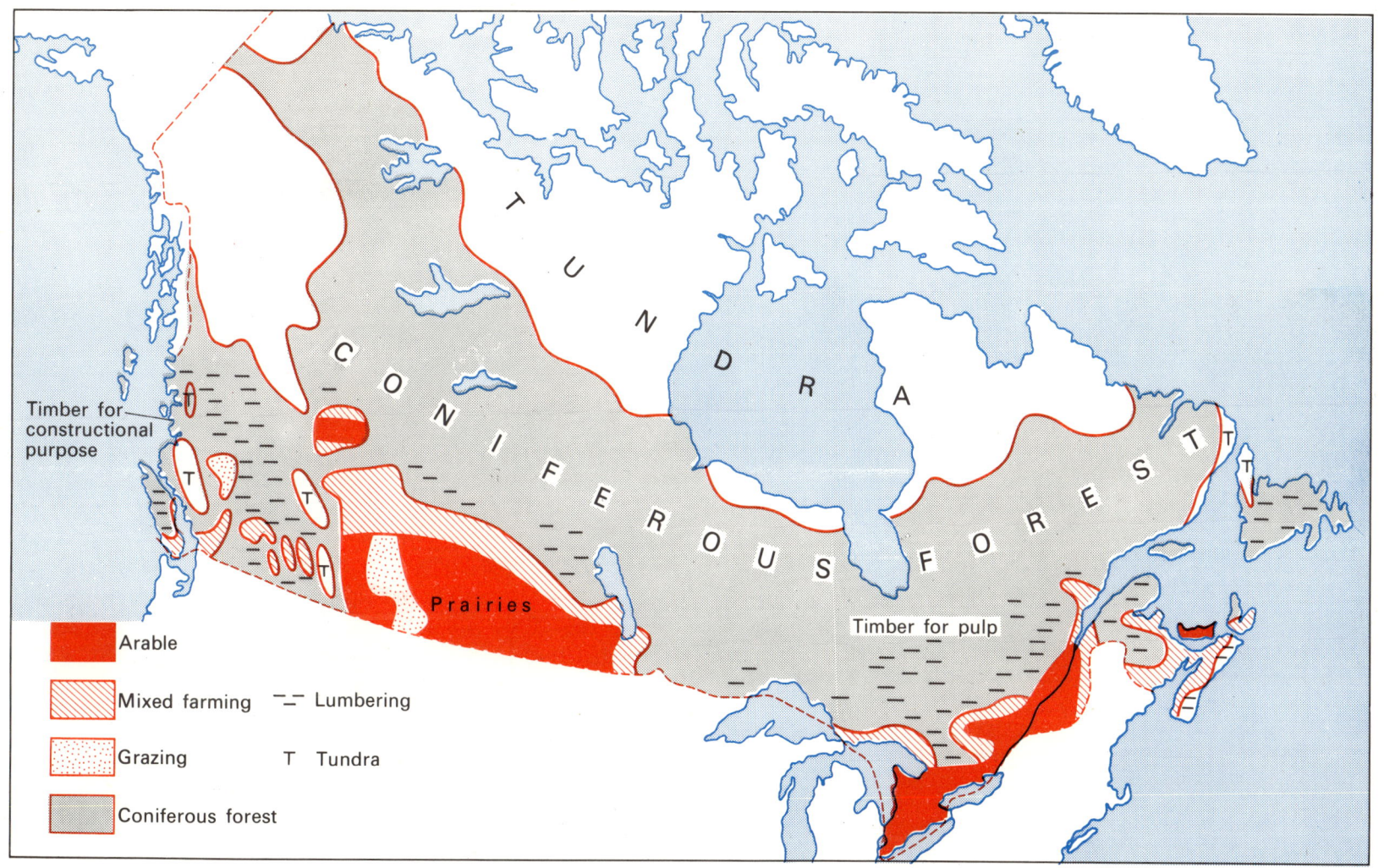

Timber for
constructional
purpose

Prairies

Timber for pulp

Arable

Mixed farming – – **Lumbering**

Grazing T **Tundra**

Coniferous forest

T U N D R A

C O N I F E R O U S   F O R E S T

FIG. 2.4 **Farming and forest resources of Canada**

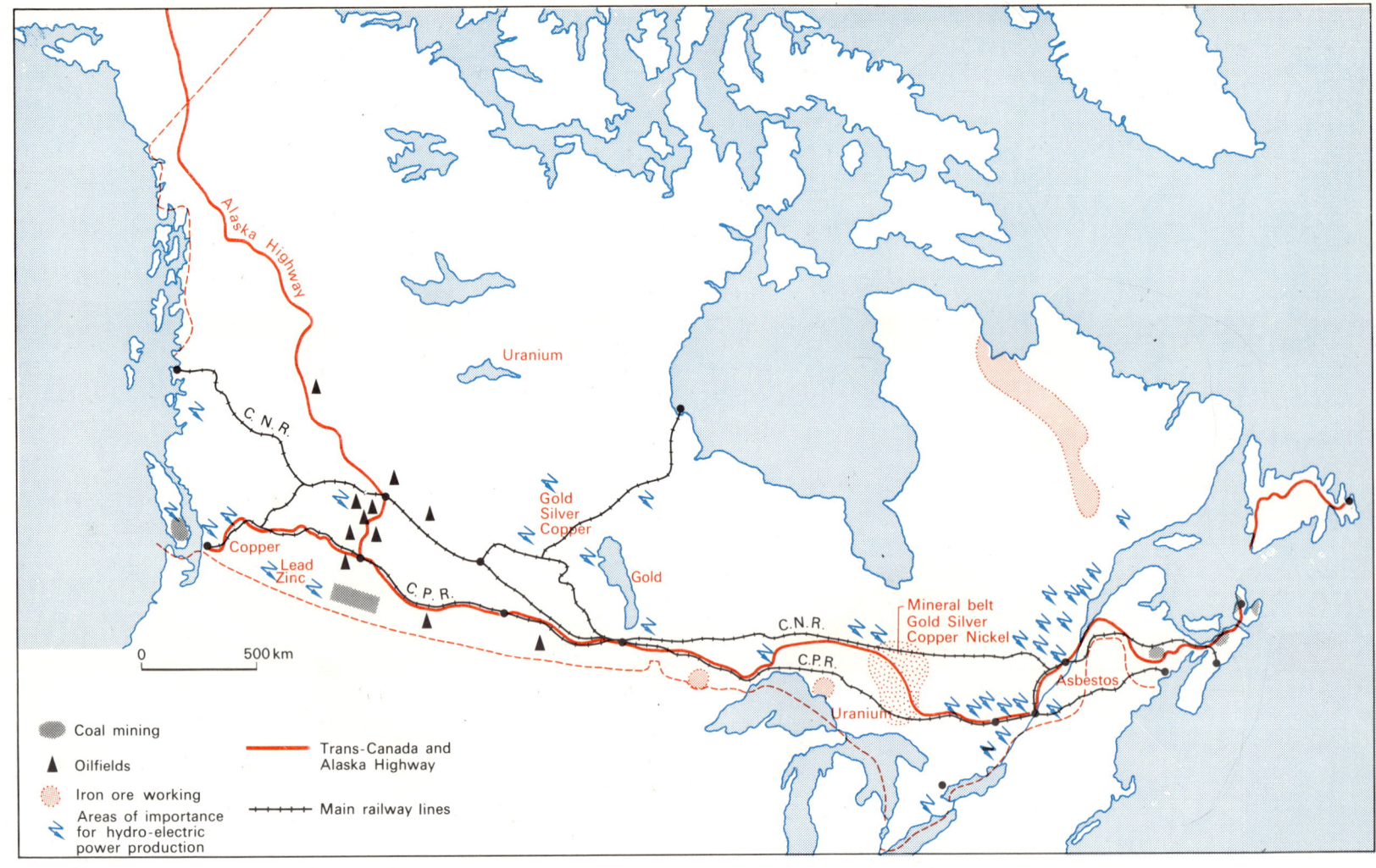

Alaska Highway

C. N. R.

Uranium

Copper

Lead
Zinc

C. P. R.

Gold
Silver
Copper

Gold

Uranium

Mineral belt
Gold Silver
Copper Nickel

C.N.R.

C.P.R.

Asbestos

0   500 km

Coal mining

Oilfields

Iron ore working

Areas of importance for hydro-electric power production

Trans-Canada and Alaska Highway

Main railway lines

FIG. 2.5 **Canada – minerals and power supplies**

Minerals and Power Supplies

Canada's prosperity owes much to its mineral wealth, and prospecting in the more remote areas is still leading to new discoveries. It is the world's leading producer of nickel, most of which comes from Sudbury in Ontario, and asbestos, from the Thetford area of Quebec. It is also a leading producer of zinc, platinum, gold and uranium. The latter, which is mined at several places in the Canadian Shield, is the chief raw material for nuclear power stations, the first of which was built at Douglas Point, Ontario. Experience gained here has led to the building of a much larger one at Pickering, near Toronto.

Large quantities of high grade iron ore are mined at Steep Rock and other parts of Ontario, and along the Labrador–Quebec border. There are vast reserves of coal in the Western Cordillera and some in Nova Scotia, but coal production has declined in recent years owing to the development of other sources of power, including mineral oil and natural gas: both the latter occur along the eastern slopes of the Rockies and in the Prairies. Oil production has increased rapidly in recent years, the principal fields being the Redwater Field north of Edmonton, the Leduc and Pembina Fields south-west of Edmonton, and a smaller field in the Turner Valley south-west of Calgary. Large quantities of oil also exist in the tar sands of the Athabasca Valley and could become of great importance in the future. Both the oil and natural gas are conveyed by pipeline to eastern and western Canada.

Water power is of great significance also, especially on the south-eastern edge of the Canadian Shield, in the Great Lakes region, and in British Columbia, where lakes, waterfalls and deep valleys provide ideal conditions for generating hydro-electric power. This will be even more important in the future, since little more than a quarter of the hydro-electric potential has so far been tapped.

Manufacturing

Canada's abundant power supplies and great wealth of raw materials have stimulated the development of many types of manufacturing industry. Until this century these were mainly concerned with the preparation of food (for example flour milling, meat packing, fish processing, and butter and cheese making), and the processing of raw materials (the smelting of metals, saw milling, and the pulp and paper industry). Today, in addition to these, there is a modern steel industry, with the main centres of production at Hamilton and Sault Ste Marie (both in Ontario), and at Sydney on Cape Breton Island. There is a large aluminium industry, based on the abundant hydro-electric power and imported bauxite or alumina, and a wide range of engineering industries including the manufacture of motor vehicles, aircraft and farm machinery. Close to the oil refineries are petro-chemical industries, producing fertilisers, synthetic rubber, fibres, polythene and detergents. Textile and clothing industries are widespread, employing, in the main, imported raw materials.

The chief manufacturing region is the Great Lakes–St Lawrence area in the provinces of Ontario and Quebec, with Montreal and Toronto as the main centres, but there are also considerable industrial developments in the Vancouver area, in the larger Prairie cities, and in some of the smaller centres along the Atlantic seaboard.

There has been much useful co-operation between the Canadian and United States Governments in major construction schemes. The Niagara Falls hydro-electric power station, the St Lawrence Seaway, and the Columbia Valley Project are well-known examples. In addition, many United States firms have established factories or subsidiary companies in Canada, producing goods for both the Canadian and United States markets.

Too much dependence on the United States has been the cause of certain weaknesses in the Canadian economy, especially in the heavy, basic industries which require great capital investment and highly skilled labour. The steel industry, however, has increased its production to over 10 million tonnes per annum which, in proportion to population, is not far behind that of the United States.

Communications

An essential factor in the development of natural resources has been the improvement of communications. In eastern Canada the inland waterways of the Great Lakes and St Lawrence provide cheap transport for bulky, heavy goods, although the winter freeze-up is a considerable handicap. The deepening of the St Lawrence between Montreal and Lake Ontario (the St Lawrence Seaway) now allows ocean-going vessels to reach the Great Lakes.

There are two main trans-continental railways – the Canadian Pacific Railway, and the state-owned Canadian National Railway. Both pass through the more populated southern regions, but there are branches leading north to less developed areas, including Churchill on Hudson Bay, Fort McMurray on the Athabasca, and the Peace River District. The interior and west of Canada depend to a large extent on the railways to carry their farm produce and minerals to the Great Lakes and coast for export.

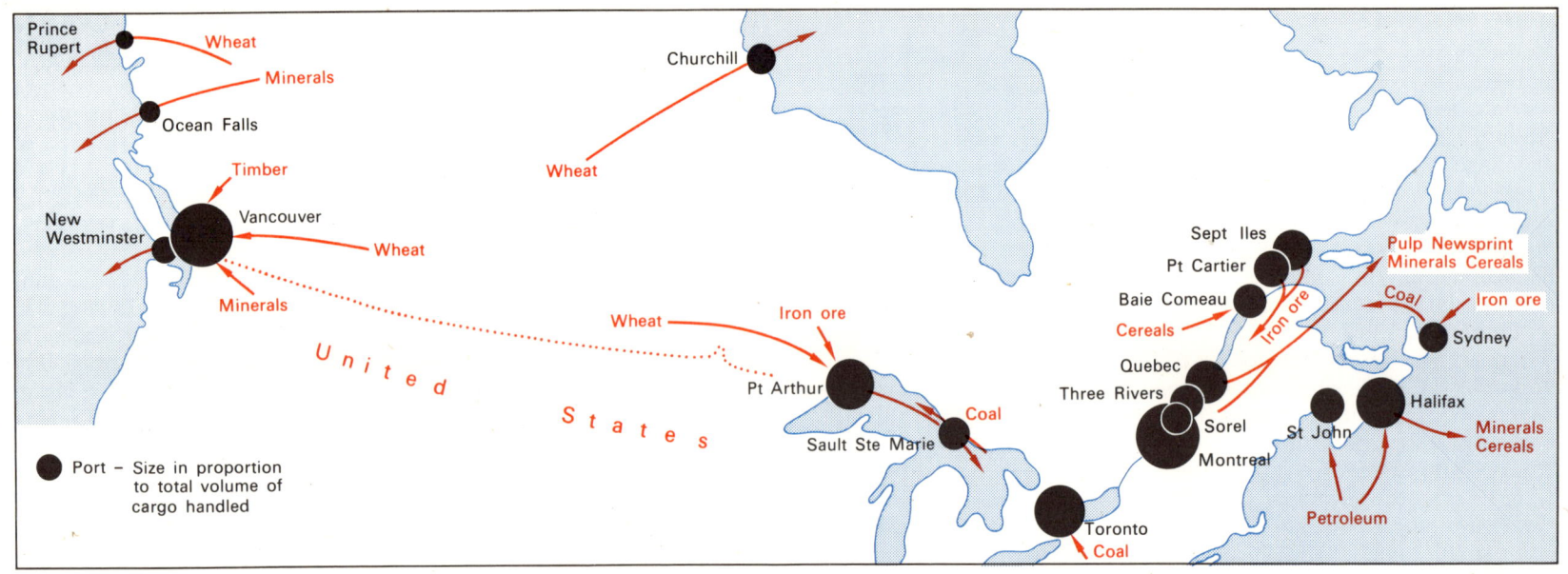

FIG. 2.6 **Major ports and trade movements**

Road transport is now competing successfully with the railways, and the road system, which is already dense in well-populated areas, is being constantly extended. In recent years the most outstanding achievement has been the construction of the Trans-Canada Highway from St John's (Newfoundland) to Victoria (British Columbia), with car ferries to bridge the sea crossings. The building of the road presented many engineering problems, and necessitated the filling in of swamps and the diversion of the Assiniboine River west of Winnipeg. In the Rocky Mountains many kilometres of snowsheds were constructed to protect the road from avalanches.

There are regular air services between the main towns and cities. Feeder services operate to more remote areas where, in addition to passengers, essential supplies are carried by air, and valuable cargoes such as furs and mineral ores are sometimes brought out.

Foreign Trade

Canada is a great trading nation. Its exports consist mainly of food, raw materials and semi-processed goods. Its chief imports are machinery, manufactured goods, and some foodstuffs and raw materials which cannot be produced at home. About 60 per cent of its trade is with the United States, 15 per cent with Great Britain, and the rest mainly with other west European countries and Japan. The value of the imports is greater than that of the exports, and the deficit is covered by an inflow of capital (i.e. investments by overseas firms and individuals) which is used to expand local industries.

Exercises

Answer in note form:
1. Explain the working of the federal system of government in Canada.
2. Where are the most valuable forests in Canada and what uses are made of them?
3. Describe in detail the route followed by the Canadian Pacific Railway.
4. Name a densely populated and a thinly populated part of Canada, and give reasons for the difference in density.
Essay:
   How have geographical factors influenced the development of manufacturing industries in Canada?

3: NEWFOUNDLAND

The province of Newfoundland consists of the island of that name and the territory of Labrador on the mainland. It was formerly a self-governing colony, but its economy did not prosper and it had to be helped by subsidies from the United Kingdom. Its present status was determined by a referendum, held in 1949, when its people chose by a narrow majority to join the Dominion of Canada. The standard of living, which was low before the war, is now improving steadily, but is still below that of other parts of Canada.

The Island of Newfoundland

The island of Newfoundland is almost as large as England, but its population numbers only 500 000. It lies across the entrance to the Gulf of St Lawrence and is a detached fragment of the Appalachian mountain system. A number of rocky ridges, which reach a height of over 600 metres in the west, cross the island from south-west to north-east and end in rugged peninsulas. There is widespread evidence of glaciation, including surfaces scraped bare of soil, rock basins scooped out by ice and now occupied by lakes, and morainic deposits of boulders and sand. Over much of the island the drainage is poor, with extensive marshes and slow-moving rivers flowing from lake to lake.

Post-glacial submergence has produced an indented coastline with many fiords, especially in the south-west, and numerous offshore islands. There is a broad continental shelf extending for about 500 kilometres into the Atlantic to the south-east, and here are the Grand Banks, which have been built up by material deposited from melting icebergs carried there by the Labrador Current.

The climate of the island is of cool temperate eastern margin type, with cold winters, cool summers, and heavy precipitation throughout the year. As an example, St John's has a mean February temperature of −6°C, an August mean of 15°C, and an annual precipitation of 1360 millimetres. Temperatures are kept low because of the influence

FIG. 3.1 **Newfoundland**

of the cold Labrador Current, and because in winter it is dominated by cold winds from the interior of Canada. The seas off the east coast and within the Gulf of St Lawrence freeze for several months, but the south coast, which receives some influence from the warm Gulf Stream, has slightly higher temperatures and its harbours are ice-free in winter. The meeting of warm and cold ocean currents in the vicinity of Newfoundland gives rise to persistent fog which, together

National Film Board of Canada

Typical scenes in the interior and along the coast of Newfoundland

with the danger from icebergs in spring and summer, makes navigation somewhat hazardous.

About half the island is forested, the chief trees being spruce, balsam fir, birch and pine, but although the heavy rainfall promotes rapid growth, the trees are not as tall as in other parts of eastern Canada. On the higher ground there is a tundra-type vegetation of lichens, reindeer moss and bilberries, as well as much bare rock.

Fishing

Fishing from the many sheltered harbours has been important for two hundred years and remains the chief occupation of the island. For reasons already mentioned (page 16) one of the world's richest fishing grounds lies off the Newfoundland coast, especially to the south-east.

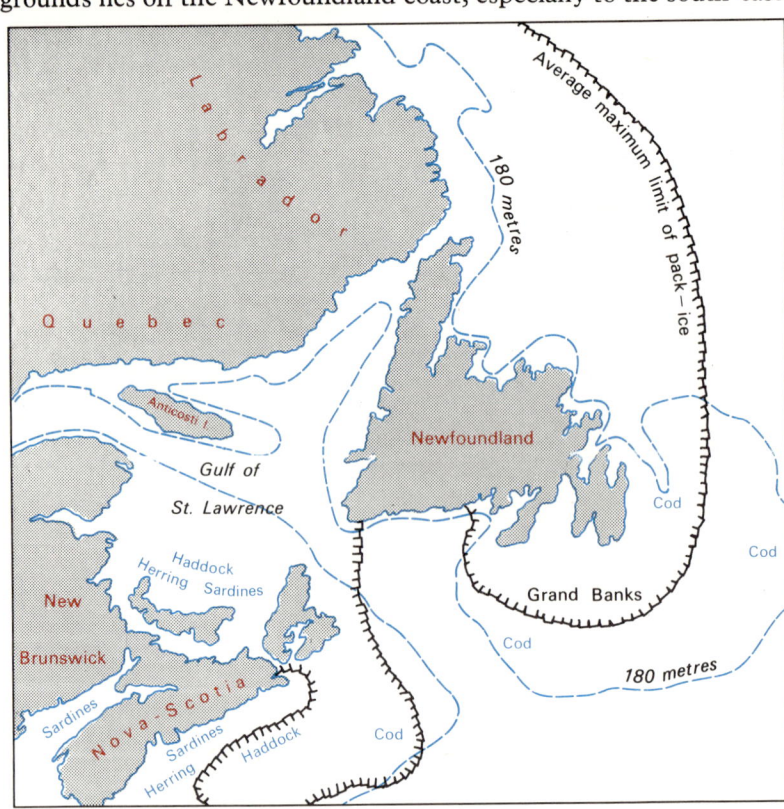

FIG. 3.2 **The coastal waters of Eastern Canada**

National Film Board of Canada
Dorymen returning to their schooner on the Grand Banks

Canadian Government Travel Bureau
Fishing fleet in Hobbs Cove, Newfoundland
Note the fish processing sheds and drying platforms.

This includes the Grand Banks, where the principal catch is cod, followed by sole, haddock, halibut, hake, mackerel and plaice. The traditional method of fishing involves the use of schooners as mother ships, with a number of flat-bottomed 'dories' working from each. Long lines, to which are attached baited hooks, are paid out from the dories. In recent years the industry has been considerably modernised and diesel-powered trawlers, fitted with refrigerated chambers, are replacing the schooners. Whereas in the past large quantities of cod were salted and dried for export to the West Indies and Latin American countries, today most of the fish is frozen for consumption in Canada and the United States.

Inshore fishing is carried on close to the shore from small boats, and accounts for about half the total catch, including herrings, lobsters, crabs and salmon. Seals are caught in spring among the ice floes of Belle Isle Strait and along the coast of Labrador, but over-killing has greatly reduced their numbers.

Forest Industries

Lumbering is carried on widely throughout the island, one of the most important trees being spruce, which is the basis of the large pulp and paper industry. Power for the mills is obtained from the many hydro-electric power stations on the rivers and lakes.

Tractor trailer loaded with logs National Film Board of Canada

Locate Corner Brook on a simple sketch map, and give as many reasons as you can for the production of newsprint there.

The first large pulp and paper mill was completed in 1909 on behalf of Lord Northcliffe, proprietor of the London *Daily Mail*, at Grand Falls on the Exploits River. Today the largest mill is at Corner Brook, which lies on tidal water at the mouth of the Humber River, and has its own hydro-electric station on Deer Lake, about 50 kilometres away. The logs are plucked from the river, chipped and mixed with chemicals so that the fibres can be broken down into pulp. The pulp is then bleached, rolled and pressed. The greatest demand is for newsprint, and large quantities of this are exported to the United Kingdom and United States.

A small proportion of the timber is used locally in constructional work which includes the manufacture of furniture, plywood, barrels and fishing boats. Because of the lack of coal on the island, timber is also burnt for fuel.

Clearing a log jam   *National Film Board of Canada*

Mineral Working

There are scattered mineral workings on the island, the most important being the mining of copper–lead–zinc ores near Buchans in the Exploits River valley, and of asbestos in the Baie Verte district on the north coast. Iron ore was mined until 1966 on Bell Island, about 20 kilometres north-west of St John's. The ore was a high grade haematite and most of it was sent to Sydney on Cape Breton Island for smelting. With the opening up of vast new open cast iron workings in Labrador and elsewhere, the Bell Island mines became uneconomic and were forced to close.

Farming

Agriculture plays little part in the economy of Newfoundland. The rugged relief, poor soils and unfavourable climate discourage farming, apart from the growing of small amounts of hay, oats and potatoes, the keeping of cattle for local milk supplies, and poultry raising in the vicinity of the main settlements. Elsewhere, small patches of poor land are farmed as a part-time occupation by woodsmen and fishermen.

Settlement and Communications

The population of the island is widely dispersed in many small towns and villages, but nearly half live on the Avalon Peninsula in the east, mostly in coastal settlements engaged in fishing. The largest of these is *St John's* (101 000), the capital and principal port. It has a deep, ice-free and almost landlocked harbour, and is the terminus of the island's only railway which links the main towns of Grand Falls, Corner Brook and Port aux Basques (from which there is a train ferry to Sydney on Cape Breton Island). The newly-built Trans-Canada Highway also connects St John's with Port aux Basques.

Newfoundland is the easternmost part of the North American continent and because of this it has submarine telegraph and telephone cable terminals. The airport at Gander was formerly used for refuelling trans-Atlantic airliners, but as the range of aircraft has increased, it has declined in importance.

Labrador

The mainland area of Labrador has a population of only 22 000. Much of it is barren plateau country with mountains rising to over 1200 metres along the coast. It is part of the Canadian Shield and contains some valuable forest areas, but its greatest wealth is in the immense reserves of iron ore which lie close to the Quebec boundary, and the power potential of the Churchill River where a major hydro-electric project is still in the early stages of construction. The geography of Labrador is dealt with as part of the Canadian Shield in Chapter 7.

Exercises

Answer in note form:
1. What effects did the Ice Age have on Newfoundland?
2. Account for the presence off the Newfoundland coast of (a) icebergs and (b) fog.
3. Why are the Grand Banks so rich in fish, and what methods of fishing are used?
Essay:
  Why is it that Newfoundland, which is almost as large as England, has a population of only 500 000?

4: THE MARITIME PROVINCES

Nova Scotia (which includes Cape Breton Island), New Brunswick and Prince Edward Island are often referred to as the Maritime Provinces. By Canadian standards each one covers only a small area, and even if joined together they would still be by far the smallest province.

Structurally the Maritime Provinces are a northern extension of the Appalachians. Most of the land is built of old, hard, crystalline rocks and has been worn down to a series of low plateaus which reach a height of about 750 metres in northern New Brunswick. The south-west to north-east Appalachian trend can be seen in the courses of many of the rivers and in the directions of the almost parallel shores of the peninsula of Nova Scotia.

The region was heavily glaciated during the Ice Age, and much of the land surface is covered by glacial deposits, with large numbers of lakes. The post-glacial submergence has produced an indented coastline and many islands, including Cape Breton Island and Prince Edward Island. The shallow Bay of Fundy is noted for its very high tides, due to the tidal waters piling up in the narrow inlets at its head. A tidal bore develops in some of the inlets where tidal ranges of between 15 and 18 metres are common.

The climate is of cool temperate eastern margin type, but is somewhat less severe than in Newfoundland. In winter the region is exposed to cold north-west winds from the interior and mean temperatures remain below freezing point for about four months, although coastal waters are free from ice except in the north. Summers are generally cool, but are slightly warmer round the Bay of Fundy which is out of reach of the chilling effect of the Labrador Current. Precipitation is heavy and well-distributed throughout the year, ranging from 800 to 1500 millimetres per annum. Halifax, which is well exposed to Atlantic influences, has a mean January temperature of $-4\frac{1}{2}°$C, a July mean of 17°C, and an annual precipitation of 1440 millimetres.

About half the total area of the Maritime Provinces remains under forest, including almost all the high ground. Most of the trees are of coniferous type, the commonest being spruce, balsam fir, pine, cedar and hemlock, with birch as the main representative of the deciduous trees.

The Maritime Provinces have more varied resources than Newfoundland, with the result that their economies have a broader basis and are not dominated by fishing and forestry. Farming, coal mining and the steel industry all play a considerable part in the life of the region. In spite of these advantages, the total population is only $1\frac{1}{2}$ million, and the standard of living is somewhat lower than in the rest of Canada except for Newfoundland. This is partly due to the isolation of the region and its poor overland communications with the large centres of population in the St Lawrence Valley and Great Lakes area, which have discouraged the development of manufacturing industries.

Farming

Because of the cool, moist climate, there is a general emphasis on livestock farming. Whilst the ancient crystalline rocks and coarse morainic deposits do not produce fertile soils, there are some lowland areas of moderate fertility where dairying is carried on and hay, oats, barley and potatoes are grown. Other areas specialise in the production of vegetables and fruit.

The Annapolis Valley in western Nova Scotia is well-known for its apples and other fruits including strawberries. The valley has been formed in soft Triassic rocks and has fertile red soils derived from a covering of glacial drift. It runs parallel to the coast of the Bay of Fundy, from which it is separated by a narrow ridge, giving shelter from strong winds and a relatively mild climate. The orchards are mainly on the lower slopes, since cold air tends to collect on the valley floor which is used for dairying and poultry keeping. Apples, apple pulp and bottled apple juice are exported, mainly to the United Kingdom.

At the northern end of the Bay of Fundy are large areas which, in their natural state, would be flooded by the very high tides. The early settlers built dykes to keep out the sea and reclaimed the marsh, which now produces valuable crops of hay and is one of the principal dairying regions of the Maritime Provinces.

Many of the lower valleys of New Brunswick, including the St John River valley, contain rich meadows and here also the main

FIG. 4.1 **The Maritime Provinces**

Map legend:
- 0 — 100 km
- Over 300m
- Coal
- National boundary
- Iron and Steel
- Food processing
- Shipbuilding
- Provincial boundary
- Timber working Pulp and paper
- P Oil Refining

deposits, and are ideally suited to high quality seed potatoes, which are supplied to all parts of Canada. Most farms are mixed farms, with dairy cattle, pigs and poultry, and there is fur farming, especially of mink, silver fox and chinchilla.

Fishing

The methods of fishing resemble those of Newfoundland except that greater use is made of trawlers. Cod, halibut, haddock, mackerel and plaice are caught on the Grand Banks as well as in local waters, and large quantities of frozen fish fillets are sent to the main centres of population in eastern Canada and the north-eastern United States. Young herring are caught off the coast of New Brunswick and canned as 'sardines'. Highest in value is lobster, which is trapped in hundreds of rocky coves round the coast including the Bay of Fundy. Salmon are caught in the rivers.

At one time fishing craft operated from many small ports, but the present trend is for the industry to become concentrated in fewer but larger ports which are equipped to handle modern trawlers and to process their catches. Halifax, Lunenberg and Yarmouth are the most important of these, and Pictou is the main centre for the lobster fisheries.

Forest Industries

The extensive coniferous forests have led to widespread lumbering, especially in New Brunswick. Much of the finest timber has already been cut and second quality timber forms the basis of a considerable pulp and paper industry. There are numerous pulp mills along the St John and other rivers, which are used both for floating the logs and as a source of hydro-electric power. In addition, there are scattered sawmills and woodworking industries.

Mining and Manufacturing

Coal is mined in the northern part of Cape Breton Island round Sydney, both along the coast and under the sea, and also round Pictou and Cumberland in Nova Scotia. These are the only coal workings on the Atlantic coast of North America, and the cheapness of sea transport has led to considerable exports of coal to other parts of eastern Canada and even to New England. The coal is burnt in local power stations, and large quantities are made into coke for use in the Sydney steelworks. In spite of recent modernisation schemes, production has declined to below 4 million tonnes per annum, and

occupation is dairying. Farms frequently combine dairying with the production of cash crops, of which potatoes are the most important, and apples, strawberries and other fruits are widely grown.

Prince Edward Island is the smallest and most intensively farmed province of Canada. It has a gently undulating surface; nowhere more than 150 metres above sea level, and three-quarters of it is under the plough. The fertile, reddish soils are derived partly from the underlying sandstones and shales, and partly from fine-grained glacial there have been many pit closures. The chief reason for this is the competition from oil and natural gas, both of which are abundant in Canada. Hydro-electricity is another important source of energy, although the output of most power stations is small because the subdued relief restricts the fall of rivers.

In addition to coal, there are a number of other mineral workings, including lead, zinc, silver and copper in the Bathurst area of New Brunswick, and gypsum and rock salt in Nova Scotia.

The first manufacturing industries to develop in the Maritime Provinces were concerned with the processing of farm produce, timber and fish, but in 1899 a major iron and steel industry was established at Sydney, using the local coking coal and iron ore brought in from Bell Island, Newfoundland. The mines on Bell Island were closed in 1966, and most of the iron ore is now obtained from Labrador. Sydney has steel rolling mills producing rails, plates and bars, and supplies steel to shipbuilding and engineering industries in the region as well as to other parts of Canada.

The tourist industry has developed rapidly in recent years, based on the wealth of game and fish, and the fine mountain and coast scenery. Thousands of visitors from the nearby densely populated areas of Canada and the United States bring considerable revenue into the region.

Settlement

The population of the Maritime Provinces is very unevenly distributed. In Nova Scotia there is a scattered mining and farming population in the north, but the mountainous south is sparsely populated except for a line of coastal settlements. *Halifax* (230 000), the capital, lies on the east coast on a peninsula between two bays, and has an ice-free, deep-water harbour. It is the terminus of the Canadian National Railway, and its trade increases in winter when the St Lawrence is closed by ice. Its exports include wheat from the Prairies, fish, timber, minerals and local farm produce. It is the principal fishing port of the Maritime Provinces, and its main industries are shipbuilding, ship repairing, oil refining and fish processing.

In New Brunswick most of the population live in the southern valleys including the St John River valley, in which lies the provincial capital, *Fredericton* (20 000). More important is *St John* (101 000), a port at the mouth of the St John River. Its harbour is cut off from the open sea by a narrow gap through which the rise and fall of the tides

National Film Board of Canada
A farm near the St John River, New Brunswick

National Film Board of Canada
Farmlands on Prince Edward Island
Compare these two scenes and give reasons for the differences. Why is farming more important in the Maritime Provinces than in Newfoundland?

The reversing falls on the St John River near St John, N.B.   *Photographic Surveys (Quebec) Ltd*

produces the famous 'reversing falls'. Because of this, ships must enter and leave harbour between low and high tides. St John is the terminus of the Canadian Pacific Railway and, like Halifax, is ice-free and has a considerable winter trade. It has ship repairing, oil refining, pulp and paper and woodworking industries.

   Prince Edward Island is exceptional in that it is fairly uniformly settled as a result of the general fertility of its soils. The capital, *Charlottetown*, has an excellent harbour and is the main trading centre of the island.

Exercises

Answer in note form:
1. Describe the physical features and climate of the Maritime Provinces.
2. Write a short account of farming in (a) the St John River Valley, (b) the Annapolis Valley and (c) Prince Edward Island.
3. Why is Halifax the leading port of the Maritime Provinces?
Essay:
   Describe the manufacturing industries of the Maritime Provinces and explain how geographical conditions have influenced their location.

5: THE ST LAWRENCE VALLEY

Between Lake Ontario and Quebec is the fertile St Lawrence Lowland, an almost level plain 480 km long and up to 110 km in width. To the north-west the land rises steeply to the Laurentian Plateau, which forms the edge of the Canadian Shield, whilst to the south are the Adirondack Mountains and to the east, the Appalachians. Across the centre flows the St Lawrence, a great navigable highway with a width of between 1 and 16 km. Between Kingston and Montreal it falls 67 metres and its course is interrupted by a series of rapids, which have been by-passed by canals. Farther downstream the gradient becomes greatly reduced and below Quebec the river opens out to a broad estuary. Here the high ground approaches both banks, leaving little room for farming or settlement, and on the south side is the mountainous Gaspé Peninsula. The St Lawrence receives a number of important tributaries, including the Ottawa, Richelieu, St Maurice, Chaudière and Saguenay, whose valleys provide easy routeways through the mountains.

Compared with the Maritime Provinces, the climate is more subject to continental influences, with the result that it has colder winters and warmer summers. Precipitation varies between 750 and 1100 millimetres per annum and is well-distributed throughout the year, falling as snow in winter. Montreal, which lies roughly in the centre of the region, has a mean January temperature of $-10°C$, a July mean of $21°C$, and an annual precipitation of 1020 millimetres.

French farmers first settled in the St Lawrence Lowland in the seventeenth and eighteenth centuries, and laid out their holdings as long, narrow strips running at right angles to the river, with the farm houses on the river bank. This gave them, in addition to a river frontage, an area of lowland backed by forest for timber supplies and hunting. As the population increased, new strips were brought under cultivation away from the river. This field pattern remains today, and the French Canadians have retained their language, customs and way of life to a remarkable degree. The present population of the region numbers nearly five million, of whom 80 per cent are of French origin. The great majority now live in towns and find employment in manufacturing industries.

Farming

The St Lawrence Lowland is an important farming region. The soils, formed on boulder clay and alluvium, are generally fertile, and the warm summers and moderate rainfall make possible the growing of most temperate crops. Wheat is not grown on a large scale, since it can be produced more economically in the Prairies, and a system of mixed farming is in general use. The main specialisation is on dairying, and large quantities of milk are sent into the towns or made into butter and cheese. Much of the land is under fodder crops, including oats, barley and roots. Pigs and poultry are also raised, and some farmers are engaged in market gardening and apple growing. Co-operative societies have been established for processing the farm produce, especially cheese, large amounts of which are exported.

Farming is less important in the hilly border country, but even here there is dairying in forest clearings in the main valleys including the Lake St John area, and fur farming is on the increase. South of the St Lawrence, on the lower hill slopes, the tapping of maple trees in spring, in order to obtain the sap from which maple syrup and sugar are made, provides a valuable addition to the farmers' incomes.

National Film Board of Canada

French farms in the St Lawrence Lowland

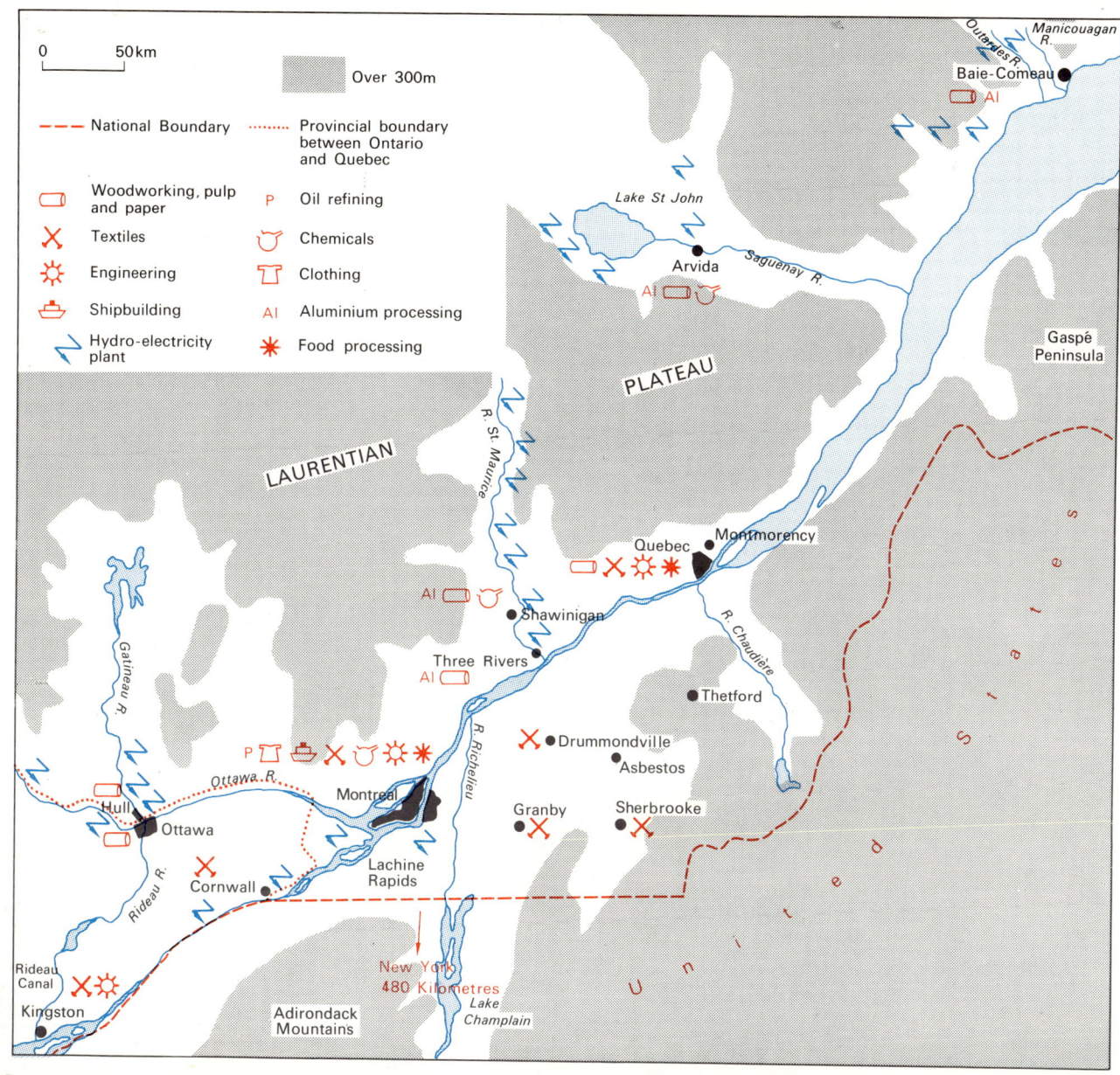

Map legend

Over 300m

National Boundary

Provincial boundary between Ontario and Quebec

Woodworking, pulp and paper

P Oil refining

Textiles

Chemicals

Engineering

Clothing

Shipbuilding

Al Aluminium processing

Hydro-electricity plant

Food processing

Map labels

Manicouagan R.

Outardes R.

Baie-Comeau Al

Lake St John

Saguenay R.

Arvida Al

Gaspé Peninsula

PLATEAU

LAURENTIAN

R. St. Maurice

Quebec Montmorency

Al Shawinigan

R. Chaudière

Three Rivers Al

Thetford

Gatineau R.

Drummondville

P Montreal

Asbestos

Ottawa R.

R. Richelieu

Hull

Ottawa

Granby Sherbrooke

Rideau R.

Cornwall

Lachine Rapids

United States

Rideau Canal

Kingston

Adirondack Mountains

New York 480 Kilometres

Lake Champlain

FIG. 5.1 **The St Lawrence Lowlands**

The farms are not as highly mechanised as in the Prairies or southern Ontario, and some are too small to provide an adequate standard of living for the farmer and his family. Nevertheless the farmers retain a strong attachment to their land, and the best farms are managed with great efficiency. French Canadians have large families and as a result there is a steady drift of population to the towns.

Manufacturing and Mining

From the point of view of employment, manufacturing industries and mining are much more important than farming. Industrial development has been encouraged by the large population of the region, which provides both a plentiful supply of labour and a market for the products. Power is derived from hydro-electric stations on the St Lawrence and its tributaries, from coal brought in from Nova Scotia, and from oil imported from Venezuela and the Middle East. There are easy communications by water, rail and road, although the winter freeze-up which closes the St Lawrence for four months each year is a serious handicap.

Among the leading industries are the manufacture of pulp and paper and sawmilling, based on the coniferous forests that cover much of the higher ground along the edges of the St Lawrence Valley, especially the Laurentian Plateau. Much of the lumbering is done by farmers who leave their farms during the winter and live in log huts in the forest. The trees are felled, dragged over the frozen ground, and stacked on the frozen rivers. When the thaw comes in spring, the logs are floated downstream to the pulpmills and sawmills, many of which lie where the fast-flowing rivers can be harnessed for hydro-electricity. Some of the largest mills are on the Ottawa and Gatineau rivers, on the St Maurice (at Shawinigan Falls), on the Saguenay and near the mouth of the Manicouagan (at Baie Comeau). The province of Quebec produces about half of Canada's wood pulp, and most of it is exported to the United States.

The abundant supplies of cheap hydro-electricity have also led to the smelting and refining of aluminium, using bauxite imported from Guyana and Jamaica. The main centres of the aluminium industry are Arvida (in the Saguenay Valley), Baie Comeau, Three Rivers and Shawinigan Falls. Work is now going on to harness the vast power potential of the Manicouagan and Outardes rivers by the construction of nine hydro-electric dams, and when complete the scheme will be the largest in Canada.

Textile manufacturing was one of the first industries in the region.

It grew up in numerous small mills situated where rivers, containing soft water, flowed from the Laurentian Plateau, and later spread to many towns on the lowland including Sherbrooke, Drummondville, Granby, Montreal and Quebec. Products are of great variety and range from wool and cotton to rayon and synthetic fibres; the manufacture of knitwear and clothing is a feature of many of the textile towns.

About half the world's output of asbestos comes from the Thetford area, where it is worked in vast quarries. Some is processed locally into asbestos sheeting, but most is exported, mainly to the United States and United Kingdom.

Towns

Montreal (2 436 000, including suburbs) is Canada's largest city and principal commercial and industrial centre. It lies on an island in the St Lawrence, close to where the Ottawa and Richelieu tributaries join

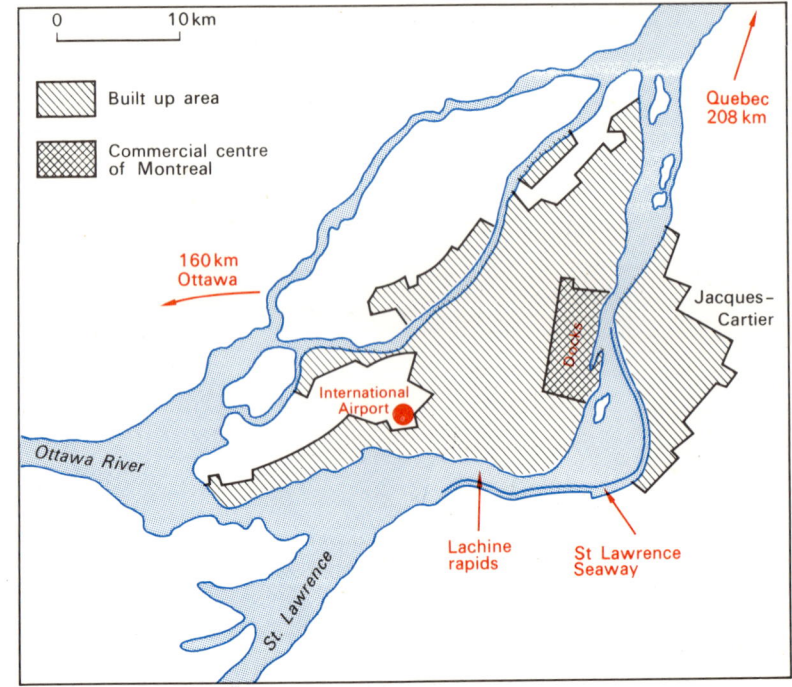

FIG. 5.2 **Montreal**

National Film Board of Canada

Montreal (above) and Quebec (right)

Compare the sites of these two cities. Why did Montreal develop into a larger city than Quebec?

the river. The island makes it a good bridging point, and the nearby Lachine Rapids were the upper limit of navigation before they were by-passed by the Lachine Canal. Montreal occupies a nodal position at the meeting point of routes (a) from the St Lawrence Estuary, (b) from the Great Lakes via the St Lawrence, (c) from the Prairies via the Ottawa Valley, and (d) from New York via the Hudson and Richelieu Valleys.

As a port it is the highest point that can be reached by large liners, so that many goods have to be trans-shipped into lake steamers. It handles about a quarter of Canada's overseas trade, exporting wheat, dairy produce, pulp and paper, and metal ores, and importing a wide range of manufactured goods and certain raw materials which are not available in Canada. Its industries, which receive power from hydro-electric stations along the St Lawrence, are of great variety and include the manufacture of textiles, clothing, chemicals and aircraft, shipbuilding and oil refining (using oil imported by pipeline from Portland, Maine).

Quite different from the cosmopolitan city of Montreal is the old French city of *Quebec* (413 000), with its narrow, cobbled streets. It grew up on an easily-defended site above a steep cliff on the north bank of the St Lawrence, overlooking the small St Charles River which provided a sheltered anchorage for ships, and is the lowest bridging point of the St Lawrence which narrows here at the head of its estuary. Quebec has developed into a major port and industrial centre, and is the capital of the province of Quebec. Its chief manufactures are textiles, leather, pulp and paper and machinery, and it obtains hydro-electric power from the Montmorency Falls and from the Lake St John area.

Ottawa (495 000) lies at the head of navigation of the Ottawa River and where it receives two tributaries, the Rideau from the south and Gatineau from the north. It is the dominion capital and has impressive parliament buildings standing on a crag overlooking the river. Coniferous forests, which cover the surrounding hills, provide timber for pulpmills and sawmills, and hydro-electric power is obtained from the local Chaudière Falls and the Gatineau River. One reason for the choice of Ottawa as capital was that it lay roughly between the English and French-speaking parts of Canada. The chief railways linking the St Lawrence Valley with the Prairies pass through it, and the Rideau Canal connects it with Kingston on Lake Ontario, though this is little used today.

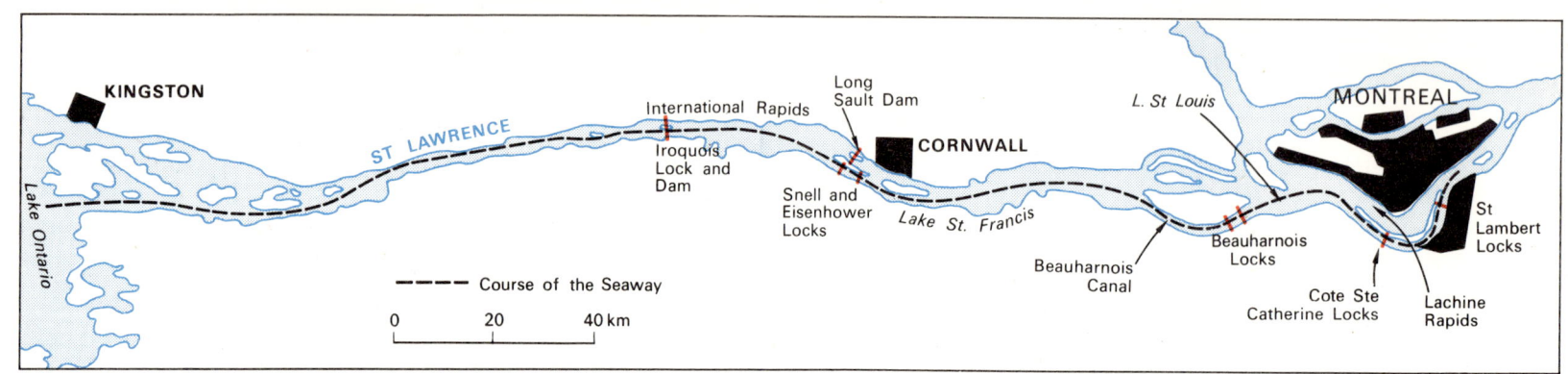

The Lockwood Survey Corporation Ltd, Toronto
The first ship to enter the St Lawrence Seaway near Montreal, after the opening ceremony in 1959

FIG. 5.3 **The St Lawrence Seaway**

The Great Lakes and St Lawrence Waterway

The St Lawrence and Great Lakes form one of the busiest inland waterway systems in the world, and provide continuous navigation for ocean-going ships from the Gulf of St Lawrence to the lakeside ports of Canada and the United States.

Navigation was formerly handicapped by the presence of rapids and waterfalls between the lakes, due to their different levels, and also by rapids along the St Lawrence between Kingston and Montreal. Canals were first constructed over a century ago and have been much improved since. The Sault Ste Marie rapids between Lakes Superior and Huron were by-passed by the Soo Canals, which contain five sets of locks built side by side. Between Lakes Huron and Erie there was a different problem in the form of a shallow delta where the St Clair River enters Lake St Clair, and this was overcome by the dredging of a short stretch of canal. The greatest difficulty was the Niagara River, which drops 99 metres between Lakes Erie and Ontario, including the Niagara Falls. To avoid it, the 45 kilometre-long Welland Canal was constructed with five sets of locks, all of which have now been twinned so that ships can pass in both directions at the same time.

Below Lake Ontario the St Lawrence passes over the International Rapids to Lake St Francis, then over more rapids to Lake St Louis, and finally over the Lachine Rapids just before reaching Montreal. A number of small canals had been built to avoid these rapids, but they formed a serious bottleneck since they could only accommodate vessels of up to 2500 tonnes thus preventing big lake boats from reaching Montreal and keeping ocean-going ships out of the Great Lakes.

The St Lawrence Seaway

A new, deep waterway along the St Lawrence had been discussed for many years and at last in 1954 the Canadian and United States governments agreed to co-operate in the project. As a result, the St Lawrence Seaway was completed in 1959, with a series of deep canals in place of the shallow ones, and seven giant locks, so that ships of 10 000 tonnes can now enter the Great Lakes.

An equally important part of the scheme was the construction of two major hydro-electric stations along the waterway, which should help to relieve the growing power shortage of the region. These are the Beauharnois Power Station 32 kilometres upstream from Montreal, and the Moses Saunders Power Station near Cornwall, whose output

is shared between the United States and Canada.

It had been hoped that the Seaway would pay for itself within 50 years through tolls levied on shipping, but so far receipts have not come up to expectations. However, its main object has been achieved, which is the provision of low cost transport for such bulky commodities as grain, iron ore, timber, coal and oil, and this should stimulate industrial development in both the Great Lakes and St Lawrence regions.

Exercises

Answer in note form :
1. Compare the climates of Montreal and Halifax, and give reasons for the differences.
2. Describe a typical French Canadian farm on the St Lawrence Lowland.
3. How were navigation difficulties overcome on the Great Lakes and St Lawrence?
Essay :
To what extent has the physical geography of the St Lawrence Lowland influenced the occupations?

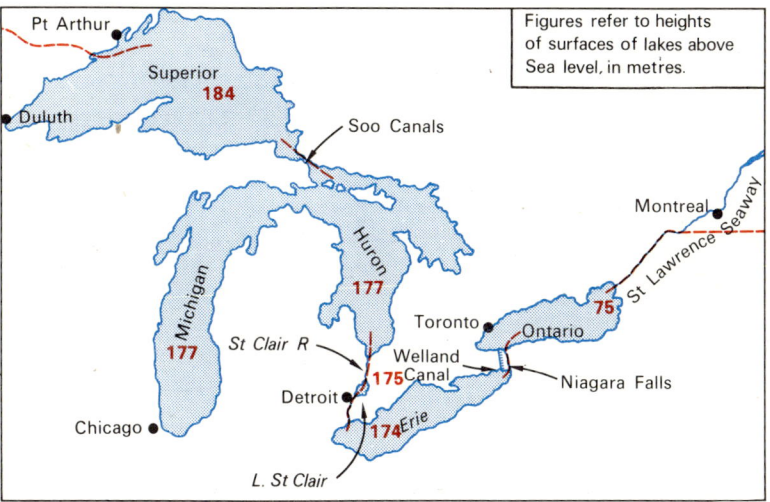

FIG. 5.4 **The Great Lakes**

6: THE LAKES PENINSULA

This small, triangular-shaped region in southern Ontario lies between Lakes Huron, Erie and Ontario, and is bounded to the north-east by the edge of the Canadian Shield. It is undulating country, built of sedimentary rocks including the resistant Niagara Limestone which forms a ridge running from Niagara Falls to the Georgian Bay, with a prominent north and east-facing scarp. The Niagara River drops 99 metres between Lakes Erie and Ontario, half of it at the Niagara Falls where it tumbles over the Niagara Escarpment into an eleven kilometre-long gorge which it has cut in the limestone.

Almost all the land has a covering of glacial material and in the central part of the peninsula there are numerous morainic ridges and oval-shaped mounds of boulder clay (drumlins). Much of the lower land along the margins of the lakes is covered by fine clays and sands, laid down in shallow water towards the end of the Ice Age when the lakes were larger than now.

The Lakes Peninsula is the most southerly part of Canada, and as a result the summers are long and warm and the winters are milder than in other parts of the Canadian interior. The lakes usually remain open water throughout the winter, although ice forms along the shores. Places near the water experience some moderating effect, giving warmer nights and slightly cooler days especially in summer and autumn. Precipitation averages between 750 and 1000 millimetres per annum and, as in eastern Canada, there is no dry season. Toronto, which lies on the shore of Lake Ontario, has a mean January temperature of $-5\frac{1}{2}°C$, a July mean of $21°C$, and an annual precipitation of 790 millimetres.

About six million people live in southern Ontario, nearly a third of the total population of Canada. The great majority live in urban areas and are employed in manufacturing industries. The region is responsible for about half the country's industrial output and has a highly developed agriculture.

Farming

Almost the whole of the Lakes Peninsula is suitable for farming, and the long growing season, large local market and fertile clays and sands have all contributed to a prosperous agriculture.

The region was formerly important for wheat growing, but after the opening up of the Prairies, most farmers turned to livestock and to the production of specialised crops. Dairy farming, as part of a mixed farming system, is carried on throughout southern Ontario. Oats, barley and lucerne are grown for fodder, and winter wheat, maize and sugar beet are the chief cash crops. Pigs and poultry are also kept on many mixed farms.

Along the shores of Lakes Ontario and Erie the mild climate and fine-grained soils have led to the growing of fruit and vegetables. The most favourable conditions for fruit are found in the Niagara fruit belt between the Niagara Escarpment and the shore of Lake Ontario. This area has exceptionally mild winters and the sandy loam soils have good natural drainage. Peaches and grapes do particularly well, and cherries, plums, apples, pears and bush fruits are also grown. The most important vegetable growing area lies to the south of Lake St Clair. It has the lowest rainfall (700 millimetres) and longest frost-free period in south Ontario, and its crops of lettuce, tomatoes, cucumbers and cabbage can be marketed earlier than those from other districts, so that they fetch high prices.

Most of Canada's tobacco is grown in Ontario, especially in a small area on the northern shore of Lake Erie, south-east of London. The soils here are a fine sand which warms up quickly in spring and the industry is highly organised, with drying kilns for the curing process.

Farms throughout the Lakes Peninsula are fully mechanised, and the number of agricultural workers is tending to decrease year by year. At the same time, farm output is constantly rising with the use of more intensive, scientific methods.

Manufacturing and Towns

It is as a manufacturing region that the Lakes Peninsula is best known. Among the main advantages for industry are: (a) the Great Lakes region of Canada and the United States is one of the most densely populated in the world and provides a vast market for manufactured goods; (b) there are abundant supplies of cheap electricity generated at hydro-electric stations on the Niagara River and along the St Lawrence; (c) the Great Lakes and St Lawrence Seaway provide cheap transport for bulky, heavy produce; and (d) although raw

The Photographic Survey Corporation Ltd, Toronto
The Hamilton steelworks
What do you understand by an integrated steelworks? Draw a sketch map to show the location of Hamilton, and indicate on the map the sources of supply of iron ore, coal and limestone.

Canadian Government Travel Bureau
Toronto and its harbour on Lake Ontario

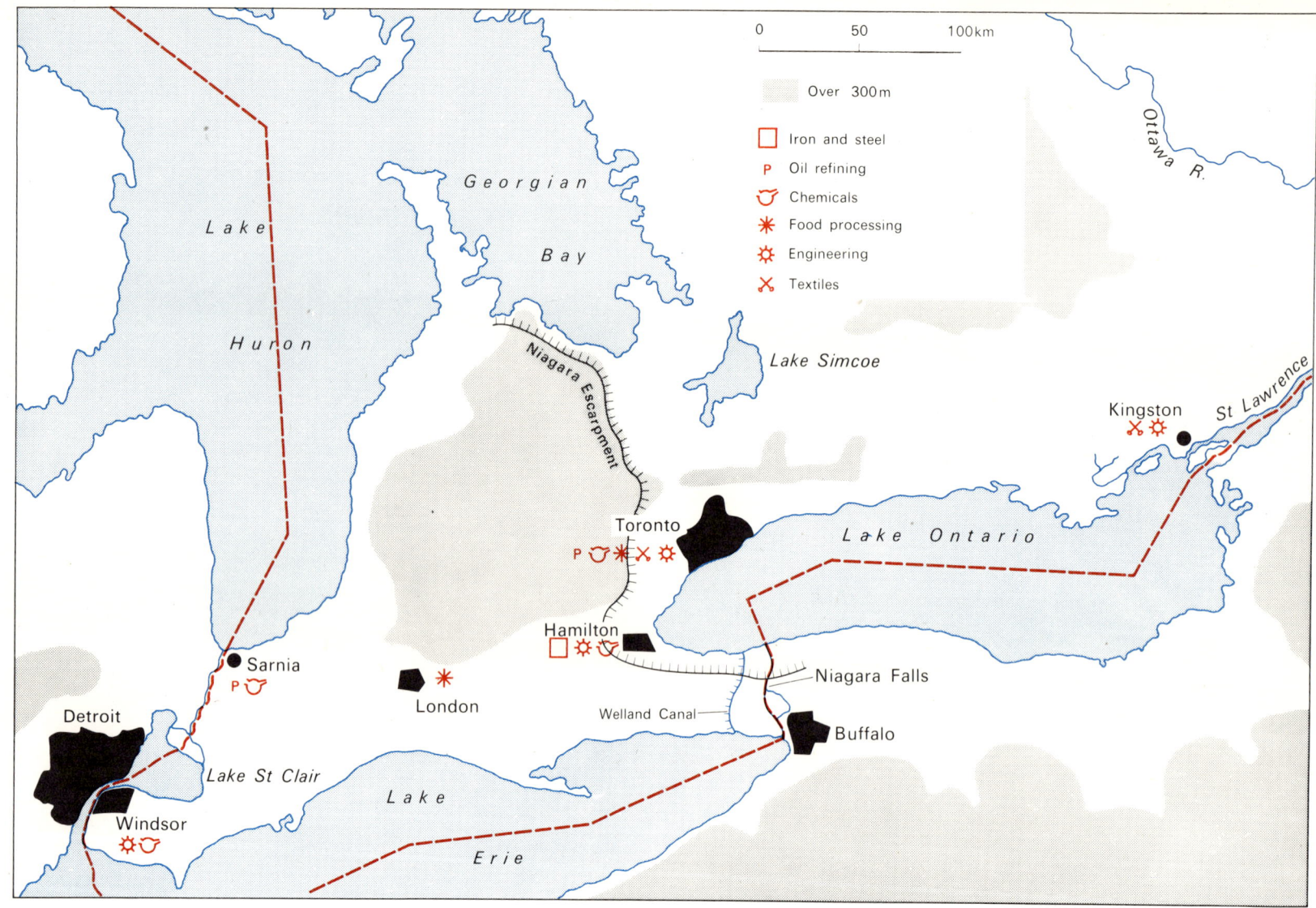

FIG. 6.1 **The Lakes Peninsula**

materials are generally lacking, there is some timber in the north, a small oilfield in the south-west, and deposits of rock salt in the Windsor area.

The industries of the region are of great variety and range from the manufacture of iron and steel, machinery, motor vehicles and aircraft to textiles, clothing, pulp and paper, furniture, and the processing of farm produce including butter and cheese-making, and the canning and quick-freezing of fruit and vegetables.

The nearness of the great American industrial cities on the other side of the lakes is sometimes regarded as a threat to Canadian industry, since they produce a similar range of goods. However, Canadian goods are protected by a tariff which raises the price of imported goods, and this has encouraged many American firms to establish branch factories in Ontario.

The further development of industry will require even greater power supplies and these will depend increasingly on thermal sources (oil, natural gas and coal), since the hydro-electric sources are almost fully developed. In addition, Canada's first nuclear power stations have been built in this region, one at Douglas Point on Lake Huron and a newer one at Pickering, east of Toronto.

Toronto (2 158 000), the capital of the province of Ontario, lies on a small bay on the north shore of Lake Ontario, on the main land route from the St Lawrence to the Mississippi Valley. In its rectangular street pattern and numerous skyscrapers it is the most 'American' of Canadian cities, and as a financial, commercial and industrial centre it is second only to Montreal. It receives its power from Niagara Falls and from its own thermal power station using coal from the Appalachians and Nova Scotia. Among its main industries are the manufacture of chemicals, machinery and textiles, motor-car assembly, oil refining and food processing. In recent years the port facilities have been greatly improved to cope with the increase in trade following the completion of the St Lawrence Seaway, and there has been a rapid growth in container traffic, partly at the expense of Montreal and the American east coast ports.

Hamilton (449 000) is an important port at the western end of Lake Ontario, and lies at the meeting point of the road and rail routes which skirt the lake. It has a large iron and steel industry, producing about 6 million tonnes of steel annually out of Canada's total production of 10 million tonnes. Iron ore is obtained from Labrador and from north of Lake Superior, and coal from the Appalachians, whilst fluxing limestone is quarried locally. Hamilton's other industries

National Film Board of Canada

The Queen Elizabeth Way, passing through the lakeside fruit belt near Jordan, Ontario

include the manufacture of machinery, chemicals, and the assembly of motor vehicles.

Windsor (212 000) is situated near the western end of Lake Erie, close to the American city of Detroit, to which it is connected by bridges and tunnels. It shares its main industry, motor vehicle manufacture, with Detroit, and also has a chemical industry based on nearby deposits of rock salt.

Sarnia is the terminus of the crude oil pipeline from Alberta and is also close to Ontario's own small oilfield. It has a large oil refinery and petro-chemical industries engaged in the production of synthetic rubber, detergents and plastics.

Exercises

Answer in note form:
1. Describe the position and physical features of the Lakes Peninsula.
2. Distinguish between the general and specialised farming activities there.
3. Explain the importance of the Great Lakes–St Lawrence Waterway to the industries of the region.

Essay:
Why is it that the Lakes Peninsula, a small region with few natural resources of its own, accounts for half the manufacturing output of Canada?

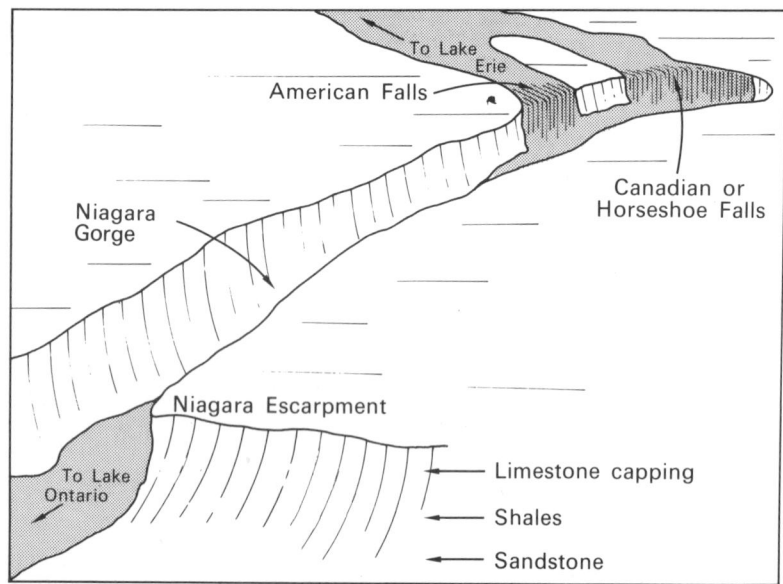

FIG. 6.2 **Niagara Falls and Gorge**

Diagram labels: To Lake Erie; American Falls; Niagara Gorge; Canadian or Horseshoe Falls; Niagara Escarpment; To Lake Ontario; Limestone capping; Shales; Sandstone

The Lockwood Survey Corporation Ltd, Toronto

Niagara Falls

Note the United States Fall (left) and Canadian Fall (above), separated by Goat Island.

National Film Board of Canada

◄ **Queenston hydro-electric plant**
There are seven large power plants in the Niagara Falls area. This one lies about six miles below the falls. Note the diversion canals which are fed from the Niagara River above the falls. Power from the area is transmitted by high tension cable to Buffalo, Toronto and a large part of New York State.

7: THE CANADIAN SHIELD AND MACKENZIE LOWLAND

The Canadian Shield is a low plateau of ancient crystalline rocks covering Labrador, the northern parts of Quebec, Ontario, Manitoba and Saskatchewan, and the North-West Territories including many islands off the coast. The Shield is saucer-shaped, with Hudson Bay occupying the lowest part and an upturned rim which is most prominent in the east and south towards the St Lawrence Valley and Great Lakes. In the west the junction of the old Shield rocks and younger rocks is marked by a number of large lakes, including Lakes Winnipeg and Athabasca and the Great Slave and Great Bear Lakes. The Mackenzie Lowland is not part of the Canadian Shield, but is a narrow northern extension of the Interior Plains, wedged in between the Western Cordillera and the edge of the Shield. In many ways its geography resembles that of the Shield, and for this reason it is included in this chapter.

During the Ice Age the whole region was covered by a thick ice sheet which removed loose material and exposed the underlying rock over large areas. When the ice melted it left behind widespread deposits of boulder clay, sand and gravel, and these deposits blocked many of the valleys, causing thousands of lakes and marshes to form. Today these are connected by winding streams, whose courses are frequently interrupted by rapids and waterfalls. The drainage is mainly into Hudson Bay, following the general tilt of the land, although a few rivers in the southern border areas enter the St Lawrence and Great Lakes, and in the north-west the waters of Lake Athabasca and Great Slave and Great Bear Lakes find their way into the Mackenzie, which flows north to the Beaufort Sea.

The climate over most of the Canadian Shield and Mackenzie Lowland is of cold temperate type, with long, severe winters and short, cool summers. Towards the north, in the tundra, winters last eight months during which the rivers and lakes remain frozen, with little daylight in mid-winter. On the other hand, there can be quite warm spells and long hours of sunshine in summer. Precipitation is generally light, varying from 250 to 500 millimetres per annum with a maximum in summer, but the amount increases to 750 or even 1000 millimetres towards the St Lawrence and Atlantic Ocean, which are open to maritime influences.

Over much of the region the natural vegetation has suffered little interference by man. In the south there is mixed coniferous and deciduous forest in which maple and birch grow alongside pine, cedar and hemlock. Farther north the forest consists almost entirely of coniferous trees, with spruce, pine and fir predominating, but much of it is poorly developed and is interrupted by stretches of muskeg (peat bog). Towards the Arctic Ocean the trees become stunted and widely spaced, and finally give way to tundra vegetation, with a profuse mixture of heathers, lichens, mosses, flowering plants and berry-bearing bushes such as bilberry and cranberry, which can survive beneath the snow for most of the year.

Because of the severe climate, infertile soils and poor communications, the Canadian Shield is one of the most thinly populated regions in the world. Settlement is concentrated in a few mining centres, pulp and paper manufacturing towns, and farming districts, especially in the south and east. Most of the north is uninhabited except for isolated ports and trading posts such as Inuvik near the mouth of the Mackenzie and Churchill on Hudson Bay, which is the terminus of a railway from the south. Undoubtedly it will have a larger population in the future, for it has vast untapped natural resources including a great wealth of minerals, timber, fish in the rivers and lakes, and water power.

Forest Industries

The forests of the Canadian Shield contain the largest reserves of softwood timber in North America, but much of the region is too remote for development. Lumbering is largely restricted to the more accessible areas north of the St Lawrence and Great Lakes. Here logs can be floated down the numerous south-flowing rivers including the Ottawa, St Maurice and Saguenay to the sawmills and pulpmills, which are located near to falls where hydro-electric power can be developed. Pulpwood is now more important than sawn timber, since most of the more valuable trees have been cut down.

The Canadian Government is taking active steps to conserve the forests by stricter control of felling and by large scale re-afforestation.

FIG. 7.1 **The Canadian Shield**

National Film Board of Canada
Typical scenery in the Canadian Shield, near Whitefish Falls, Ontario

National Film Board of Canada
The International Nickel Company's reduction plant at Copper Cliff, near Sudbury

In the past, great losses have resulted from forest fires and much attention is now being given to fire-prevention measures. Hundreds of observation towers have been built, and mobile fire-fighting teams are always ready for action.

A wide variety of fur-bearing animals live in the forests, and trapping is a significant occupation in spite of the continued decrease in the numbers of animals and growing competition from the fur farmers of southern Ontario and Quebec. Trapping is carried on mostly in winter when the furs are at their best and the trappers, who include many Indians and Eskimos, still travel by dog sledge and on snow shoes. The pelts are sold at the Hudson's Bay Company's trading posts, which also supply the trapper with many of his needs. The most valuable pelts are those of the Arctic fox, squirrel, muskrat, beaver and mink.

Farming

For reasons of climate and soil, most of the Shield lands are unsuitable for agriculture. Only towards the south, where the summers are somewhat warmer, is it profitable to grow such hardy crops as hay, oats, barley and potatoes, and to keep herds of dairy cattle. Two of the most important farming areas are the Cochrane Clay Belt, which was formerly the bed of a glacial lake and extends for a considerable distance along the Canadian National Railway line, and the Saguenay Valley round Lake St John, where marine silts and river alluvium provide fertile soils. But even these more favoured areas are only partially farmed, and many farmers work in the forests in winter in order to supplement their incomes.

Mineral Working

In many parts of the Shield there is evidence of igneous intrusions, i.e. of molten rock (magma) having been forced into the surface rocks. The magma, as it cooled, deposited metallic ores as veins in the rock, thus producing concentrations of valuable minerals.

The first important discoveries were made along the newly-constructed trans-continental railways in western Quebec and Ontario, and even today most of the workings are in this region. Elsewhere, development has depended on transport facilities, including the provision of railway branch lines and new roads like the Mackenzie Highway. The freezing of the rivers does not help, although in summer steamer services operate on the Mackenzie, Athabasca and Slave Rivers. Prospectors, employed by the government as well as by private companies, are still at work in many remote areas, depending on supplies brought by aeroplane (fitted with skis for landing on ice in winter) or helicopter.

Canada is the third largest gold producer in the world, after South Africa and the U.S.S.R., the output coming mainly from the Canadian Shield. Some of the principal gold mining centres are Porcupine, Kirkland Lake and Timmins in eastern Ontario, Rouyn-Noranda in western Quebec, and Yellowknife on the shores of Great Slave Lake in the North-West Territories.

Nickel-copper ores are worked in the Sudbury area of south-east Ontario, some by open cast methods but most by underground mining. In addition to nickel and copper, smaller quantities of cobalt, gold, silver, platinum and other metals are recovered during the extraction process. Copper and zinc ores are mined at Kirkland Lake, and there are a number of important mining centres in northern Manitoba including Llyn Lake (copper, silver, nickel), Flin Flon (copper, zinc) and Thompson (nickel). Canada produces about 60 per cent of the world's nickel and most of it is exported to the United States and Western Europe, where it is in demand for use in armour plate, jet engines and stainless steel.

There are large iron ore deposits to the north of Lake Superior, the chief workings being at Steep Rock and near Michipicoten, from which the ore is sent to steel works at Sault Ste Marie and Hamilton. Since the war a great new iron field has been developed on the borders of Quebec and Labrador, round Knob Lake and Wabush Lake in Labrador, and also at Lake Jeannine in Quebec. Working is open cast and power supplies, obtained from local hydro-electric stations, will be greatly increased by the completion, in 1972, of the Churchill Falls generating station, which will be one of the most powerful in the world. The Knob Lake ores are high grade, with an average iron content of 55 per cent, but those of Wabush Lake and Lake Jeannine are lower grade and have to be concentrated before despatch. Two railways have been built to handle the ore, one from the chief mining centre of Schefferville to Sept Iles on the Gulf of St Lawrence, and the other from Lake Jeannine to Port Cartier. Much of the ore is shipped to the United States and to Canada's own steelworks at Hamilton and Sydney, and some is also sent to the United Kingdom, Italy and Japan.

There are also large reserves of uranium, the principal raw material used in nuclear power stations. The most important workings are at Uranium City on Lake Athabasca, Port Radium on Great Bear Lake and Elliot Lake close to the Georgian Bay in Ontario. Most of the

National Film Board of Canada
The new community of Labrador City, near Wabush Lake, with the iron ore reduction plant in the background

output goes to the United States and United Kingdom, since Canada has only two nuclear power stations at present.

The Eskimos

Some 30 000 Eskimos are to be found in the North American tundra, mainly along the Alaskan and Canadian Arctic coasts including the shores of Hudson Bay and the south and east coasts of Baffin Island.

European contacts have brought great changes in the lives of many Eskimos. The introduction of money has led many of them to adopt a diet of tinned food, coffee and bread, and to give up their traditional way of life. Some have become trappers of fur-bearing animals and others have found regular employment. In an effort to help them, the Canadian and United States governments have introduced domesticated reindeer and are providing health, educational and other social services. The population, which had been in decline, is now growing again. The Eskimo has proved himself to be quick to learn and many have become skilled technicians, whilst others work as artists and sculptors.

National Film Board of Canada

Summer and winter on Baffin Island

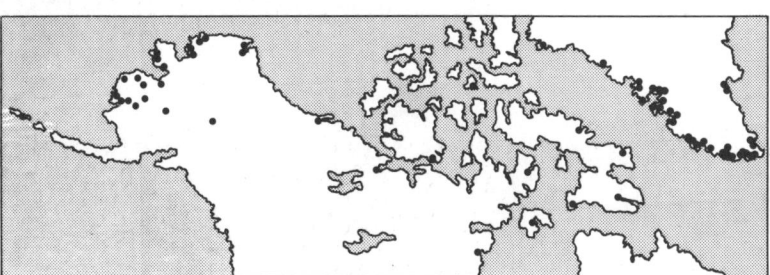

FIG 7.2 **The distribution of Eskimo groups.** Each dot represents the approximate location of an Eskimo settlement

Some Eskimos still cling to the older way of life. Living in small, scattered family groups, they have developed a remarkably organised life in difficult surroundings, based mainly on hunting and fishing. Their chief prey is the seal, which provides them with meat, blubber (a layer of fat beneath the skin yielding an edible oil and also used for cooking and lighting), skin for clothing, bones for implements and sinews for thread. Walrus and whales are also caught, and for transport they use dogs on land and kayaks in water. In winter they live in huts of turf or stone near the sea, where they can kill seals as they come up to breathe at holes in the ice. Occasionally, on hunting trips, they build temporary houses of snow known as igloos. In summer they live in tents and hunt the caribou and other animals which come north from the forests. They also fish for salmon, trout and cod, and collect wild berries, roots and edible fungi. But even among these Eskimos changes are taking place; the rifle and motor boat are replacing the harpoon and kayak, the money to buy these being obtained by the sale of surplus skins, fish and carvings in wood and stone.

Exercises

Answer in note form:
1. Describe the physical features of the Canadian Shield.
2. Compare the natural vegetation of the cold temperate region with that of the tundra.
3. What effects has contact with Europeans had on Eskimo life?
Essay:
   Write an account of the mineral wealth of the Canadian Shield and its present exploitation.

8: THE PRAIRIES

	Mean January Temperature (°C)	Mean July Temperature (°C)
Calgary (1030 metres above S.L.)	−11	16
Winnipeg (455 metres above S.L.)	−19	19

The Prairies are a vast expanse of flat country, formed of horizontally-bedded sedimentary rocks and stretching from the Ontario border to the Rocky Mountains, through southern Manitoba, Saskatchewan and Alberta. The land rises in three huge steps from about 300 metres above sea level in the east to over 1000 metres in the west. The lowest step, in Manitoba, is a level plain drained by the Red River and Assiniboine. Towards the end of the Ice Age the area was covered by a large lake, of which Lakes Winnipeg, Manitoba, Winnipegosis and Lake of the Woods are the shrunken remnants, and the silt deposited in the lake now forms dark, fertile soil. The second step lies in western Manitoba and eastern Saskatchewan, and is a rolling plateau broken by isolated hills, and the third step forms the high plains of western Saskatchewan and Alberta which extend as far as the Rocky Mountains. The Prairies are crossed by a number of large rivers, including the North and South Saskatchewan and Athabasca, whose volumes vary greatly according to season, being greatest in spring when there is a danger of flooding, and lowest in late summer.

The climate is everywhere of temperate interior type, with warm summers and very cold winters. Most areas have about five months with mean temperatures below freezing point, whilst in summer heat waves are a common occurrence. Alberta is influenced from time to time by a warm, dry wind from the Rockies called the Chinook, which blows on the south side of depressions as they pass across the region. The wind is warmed as it descends the mountains and its effect is most marked in winter. The following figures show the extremes of temperature resulting from the absence of maritime influences, and also the effect of the Chinook which gives Calgary the slightly higher winter temperatures. The cooler summers at Calgary are due to its greater altitude.

The annual precipitation is light, varying from 500 millimetres at Winnipeg to 400 millimetres at Calgary and as little as 320 millimetres in some districts of south-east Alberta. The greatest precipitation comes in late spring and summer, often in the form of thunderstorms, and there is only light snowfall in winter. An unfortunate aspect of the climate is its unreliability, with great variations in precipitation from year to year, and abrupt changes of temperature.

The Prairies are natural grasslands, and trees appear only along the watercourses and in the forest borders to the north. Large numbers of bison and other wild animals formerly ranged over the grasslands, but today almost all the land is farmed. The soils, which are derived from glacial drift rather than from the solid rock, are deep, fertile and dark in colour, with a rich concentration of humus near the surface as a result of the growth and decay of the prairie grasses over thousands of years. In the drier west, where the grasses were poorer, the humus content is lower and the soils are chestnut brown in colour.

Farming

Almost all Canada's wheat is grown in the Prairies. The first settlers used the land for cattle ranching; then after 1880 the completion of the trans-continental railways encouraged the rapid spread of wheat growing. The region had been thought of as unsuitable for wheat until a Scottish farmer, David Fife, succeeded with a hardy, quick-ripening European variety to which the name 'Red Fife' was given. Red Fife was later crossed with an Indian variety and the result was called 'Marquis'. Marquis could ripen in 110 days which made possible a considerable northern extension of wheat growing. It remained in use until about 1940 when new, improved types were developed which are able to mature in well under 110 days.

Prairie wheat is spring-sown, because of the severity of the winters, and the grain is of a hard variety which is excellent for flour milling. The conditions which favour wheat growing in the region can be summed up as follows: (a) the large expanses of flat land, suitable for mechanised farming; (b) the fertile, easily-worked soils; (c) the rainfall of between 380 and 500 millimetres, falling mostly in the growing season; and (d) warm, sunny periods during the summer for ripening and harvesting.

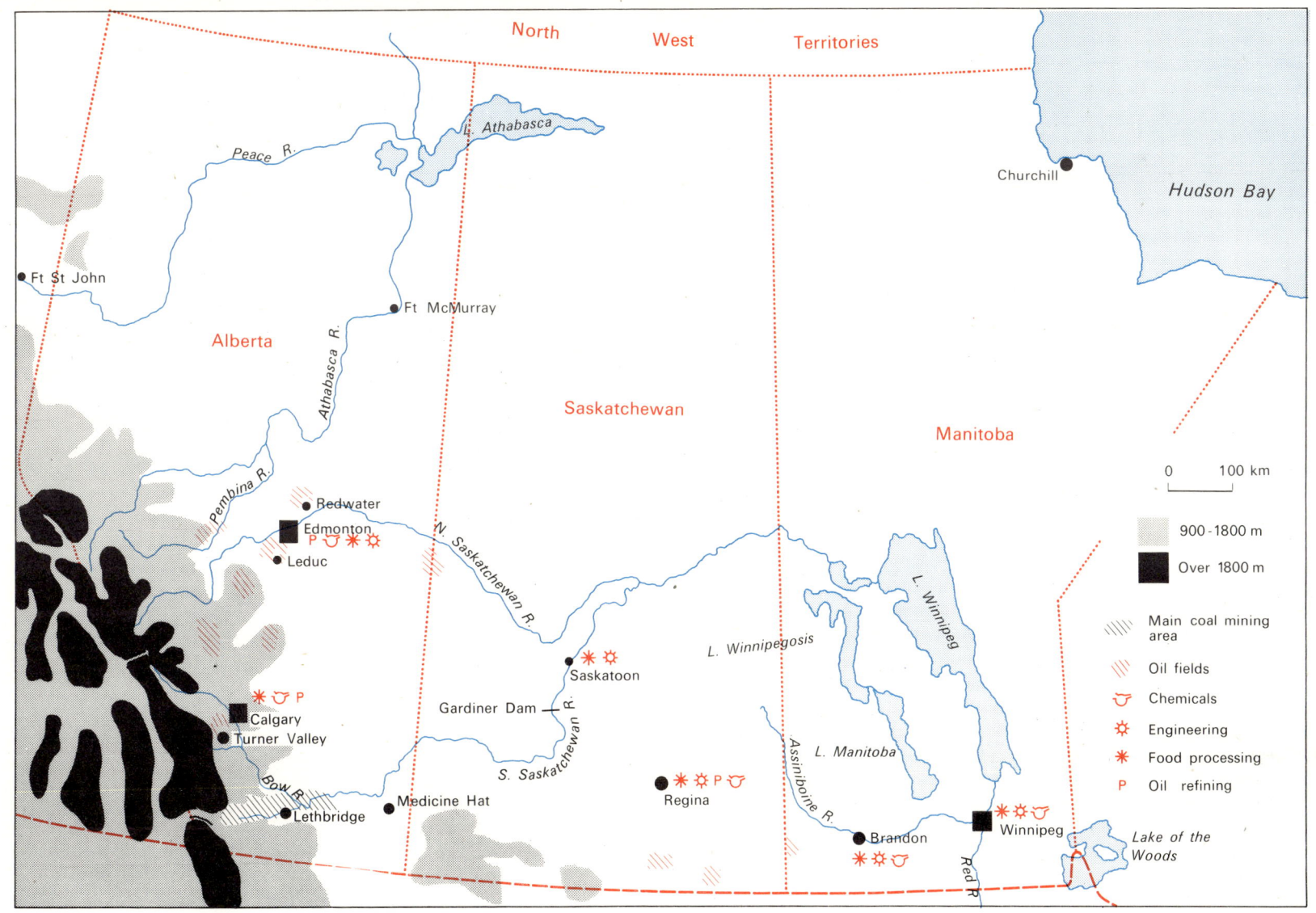

FIG. 8.1 **The Prairie Provinces**

Combines harvesting wheat near Regina, Saskatchewan

Prairie scene in Saskatchewan, with grain elevators along the railway

Many of the farms are between 400 and 800 hectares and consist of large, rectangular fields. They are highly mechanised and little labour is needed except at harvest time, so that although yields per hectare are lower than in Western Europe, output per worker is much higher. After the harvest, most of the grain is stored in elevators along the railways and is then taken by rail to the ports. The principal routes, in order of importance, are:

(1) Through the Rockies to the ice-free ports of Vancouver, New Westminster and Prince Rupert, for export to the Far East and to Europe via the Panama Canal. China and Japan are now Canada's best customers after the United Kingdom.

(2) To Fort William and Port Arthur on Lake Superior. Since the opening of the St Lawrence Seaway some grain is sent direct from these ports to Europe by ocean-going freighter, but most is still carried by lake freighter and transferred to ocean-going vessels at Montreal and other St Lawrence ports. In winter, when the St Lawrence is closed by ice, grain is sent by rail to New York, St John and Halifax.

(3) To the small Hudson Bay port of Churchill, which is ice-free from late July to early October.

Canada is the world's greatest exporter of wheat, and exports about three-quarters of its crop. The harvest varies greatly from year to year, and when there is a bumper crop prices fall drastically. To prevent hardship, farmers are protected from losses by government support.

The Prairies are not exclusively under wheat. Oats and barley are also important, especially to the north of the main wheat belt, whilst in the southern parts of Saskatchewan and Alberta, where the rainfall is low, there is large scale cattle and sheep ranching. Alberta benefits from the warming effect of the Chinook, which makes it possible for the cattle to remain out of doors in winter and reduces the need for winter feed.

Dams have been built across a number of rivers where they leave the Rockies, including the Bow River near Calgary and other tributaries of the South Saskatchewan near Lethbridge, to supply irrigation water for the growing of vegetables, fruit, sugar beet, alfalfa and maize, as well as for generating hydro-electricity. The recently completed giant Gardiner Dam on the South Saskatchewan has brought more land under intensive cultivation near Saskatoon. In some areas, where the rainfall is inadequate for ordinary cultivation, dry farming is practised. Under this system the land is left fallow every other year, being frequently harrowed so that some of the rain will pass into the subsoil.

In recent years new areas to the north of the Prairies have been opened up for farming, particularly where roads and railways exist.

Manitoba Department of Industry and Commerce

Grain elevators at Churchill

What do these three photographs tell you about the production and export of Canadian wheat?

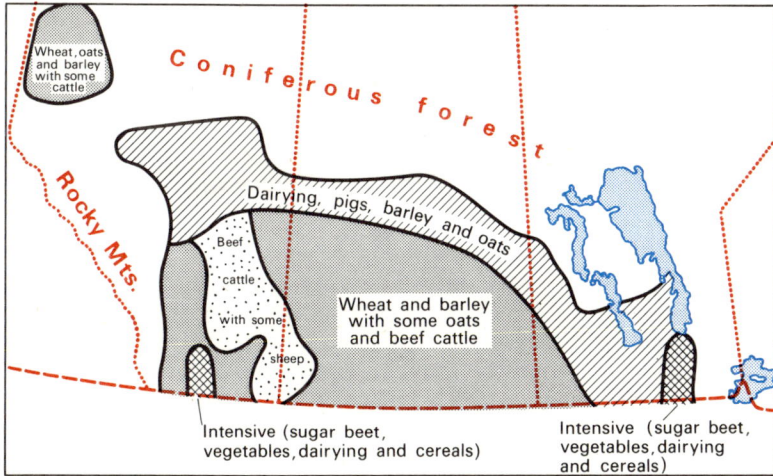

FIG. 8.2 **The Prairies.** Only a small proportion of the three Prairie Provinces is prairie. This map shows the extent and use of agricultural land which makes up the prairies

An example is the Peace River District where the low altitude and long summer days reduce the severity of the climate so that quick-maturing wheat, barley and fodder crops for cattle and pigs can be grown.

Farming on the Prairies is not without its hazards, which include late spring frosts, hailstorms, blizzards, grasshoppers and drought. The greatest problem is soil erosion, one reason for which is the growing of wheat year after year on the same land. The government is encouraging the planting of tree belts to break the force of the wind, and is advising farmers to change from pure arable farming to mixed farming. This has already been widely adopted in the eastern Prairies, where the growing of cereals, sugar beet and fodder crops is combined with cattle rearing and dairying.

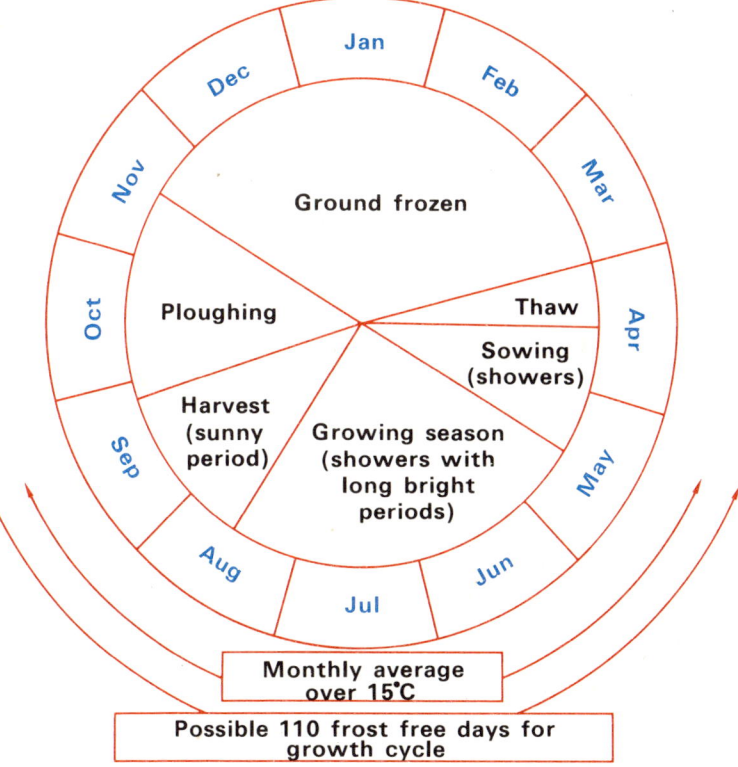

FIG. 8.3 **The farmer's year in the prairies**

Industry and Mineral Working

Until the last war the only industries of importance in the region were concerned with the processing of farm produce, together with some agricultural engineering. Today these are still carried on and in addition, a whole range of chemical industries using oil, natural gas and potash as raw materials, have made their appearance.

There are large deposits of coal in the eastern foothills of the Rockies and in the western Prairies. Sub-bituminous coal and lignite are produced at huge open cast workings especially near Calgary and in southern Saskatchewan, and bituminous coal is mined near the Crows Nest Pass and in the Edmonton area. The present production of coal amounts to about 7 million tonnes per annum and most of it is used in electricity power stations. Coal production could be much greater, but is kept down by competition from oil and natural gas, both of which are cheaper to produce and more easily transported.

Almost all of Canada's oil comes from the Prairie Provinces, the main oilfields being in Alberta, with smaller ones in Saskatchewan and Manitoba. The first oil strike was made in 1914 in the Turner Valley near Calgary, but production from this area has declined. The greatest output now is in the Edmonton area, especially from the Pembina, Redwater and Leduc fields. A pipeline connects the area with Sarnia on Lake Huron and Toronto, and another crosses the Rockies to Vancouver and north-west U.S.A. Enormous quantities of oil are known to exist in the tar sands of the Athabasca Valley north of Edmonton; production has been held back by technical difficulties, but these have now been largely overcome.

Natural gas, which occurs in southern Alberta and the Peace River district, has become increasingly important in recent years. Edmonton, Calgary, Medicine Hat and other towns use the gas both in industry and in the home, and there are pipelines carrying the gas to eastern Canada and Vancouver. Sulphur, which is used in the manufacture of wood pulp and fertilisers, is extracted from the gas, and in the Edmonton area there are chemical industries based on both oil and gas.

One of the world's largest known reserves of potash underlies part of southern Saskatchewan, near Saskatoon. It is worked at depths of between 800 and 1100 metres, and most of the output is used in the manufacture of fertilisers.

Population and Towns

In spite of a few bustling cities, the Prairie Provinces are thinly

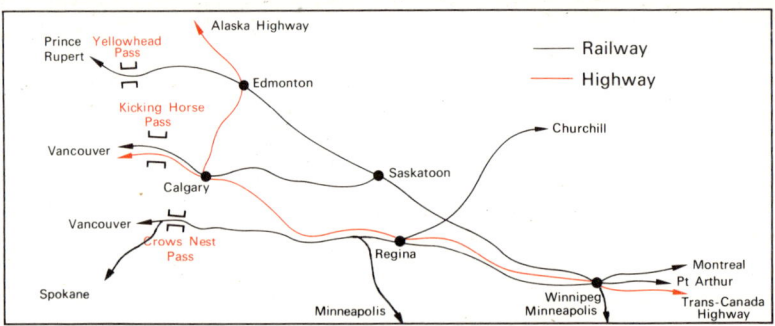

FIG. 8.4 **Communications in the Prairie Provinces**

populated. Extensive farming does not require much labour, and the typical landscape is one of many large fields but few homesteads. Of the total population of 3½ million, the great majority are town-dwellers, employed mainly in manufacturing industries and commerce.

The largest town is *Winnipeg* (510 000), the capital of Manitoba, which lies at the junction of the Red River and Assiniboine, in the middle of the fertile Red River basin. Originally it was a fur trading post, but with the coming of the railways and the opening up of the Prairies, it became the principal centre for the grain trade. Both the Canadian Pacific and Canadian National Railways run north of the Lake of the Woods and south of Lake Winnipeg, so that they are forced to pass through Winnipeg. It has many light industries serving the Prairie region, including flour milling, meat packing and the making of agricultural implements.

Regina (131 000) is the capital of Saskatchewan and an important railway junction. It serves a rich agricultural area and has food processing and engineering industries.

Calgary (330 000) lies on the Bow River in the foothills of the Rocky Mountains, and is a route centre commanding the Kicking Horse Pass route to Vancouver and another pass giving access to the upper Columbia Valley. It has food processing industries and chemical industries based on nearby coal and oil supplies. Its links with cattle ranching are reflected in its cowboy traditions which show themselves in the annual stampede, rodeo displays and waggon races.

Edmonton (438 000), Alberta's capital, lies on the northern edge of the Prairies to the east of the Yellowhead Pass. It is the gateway to the north, with railways leading to Fort St John, the Peace River district and Fort McMurray. Edmonton is the boom town of the

The Potash Company of America's plant at Patience Lake, Saskatchewan

Bucket wheel excavator at the Great Canadian Oil Sands Ltd plant at Fort McMurray, Athabasca valley, Alberta

The Rimbry Natural Gas plant in Alberta

Prairies and has doubled its population in the last ten years. This recent growth owes much to the development of oil and natural gas fields, and it has large oil refineries and petro-chemical industries producing synthetic fibres, polythene sheeting and plastics. It also has aircraft and food processing industries.

Exercises

Answer in note form:

1. What do you understand by a continental type of climate, as found in the Prairies?

2. Locate the main deposits of oil and natural gas in the Prairies, and state how they have influenced life in the region.

3. Describe the routes used for the export of wheat, and give the advantages of each route.

Essay:

Discuss the geographical reasons that have led to a specialisation in wheat growing in the Prairies.

9: BRITISH COLUMBIA AND THE YUKON

British Columbia lies almost entirely within the Western Cordillera, and has a complex structure resulting from intense folding, faulting and widespread igneous intrusions. From east to west the following parallel belts can be recognised:

(1) The Rocky Mountains, which are almost 4000 metres high and whose summits owe much of their spectacular shape to ice sculpture. The Rockies can be crossed by a number of fairly easy passes, including the Crow's Nest, Kicking Horse and Yellowhead Passes.

(2) A deep, longitudinal valley called the Rocky Mountain Trench, following a line of weakness caused by faulting and occupied by the upper courses of the Fraser, Columbia, Kootenay and Peace rivers.

(3) A more broken range of high land which includes the Selkirk, Caribou and Stikine Mountains.

(4) The much dissected British Columbian Plateau, whose average elevation is about 900 metres.

(5) The Coast Range which, although not generally as high as the Rockies, contains Mount Waddington (4042 metres), the highest peak in British Columbia. The mountains fall steeply to the coast where valleys, deepened by glaciers during the Ice Age, have been submerged to form fiords.

(6) A partly drowned mountain range forming a line of mountainous islands off the coast, including Vancouver Island and Queen Charlotte Islands. The islands are separated from the mainland by the Inner Passage, a sheltered sea route from Vancouver to Alaska.

There are many powerful rivers, fed by copious rainfall and melting snow, of which the Fraser, Columbia, Skeena and Stikine are the largest. Most have lakes in their upper courses, and follow longitudinal valleys before turning abruptly to pass through the mountain ranges.

The climate of British Columbia is of cool temperate western margin type, and is under the influence of south-westerly winds blowing off the warm waters of the North Pacific Drift. As a result, temperatures are markedly equable along the coast, with mild winters and cool summers, and harbours are ice-free throughout the year. The precipitation, falling mostly in winter, exceeds 1500 millimetres per annum in many places and reaches 5000 millimetres where high mountains rise steeply from the coast. Typical of the coastal tract is Vancouver, which has a mean January temperature of 2°C, a July mean of 17°C, and an annual precipitation of 1519 millimetres.

Beyond the Coast Range continental influences make themselves felt. The British Columbian Plateau and Rocky Mountain Trench have extremes of temperature and a precipitation of under 400 millimetres, but the precipitation again increases on the slopes of the Rocky Mountains and other high ranges.

The mild, moist climate encourages rapid tree growth, and western British Columbia carries the most luxuriant forests in Canada, with giant conifers including Douglas Fir, hemlock, red cedar and spruce. The Douglas Fir attains a height of between 35 and 55 metres, and has a base diameter of up to 4.25 metres. Its timber is uncommonly strong and durable, and is ideal for use in the building industry, in mines and on the railways. Inland, the higher slopes are also forested, but on the drier plateaus and in the deeper valleys, trees give way to grasslands.

Forest Industries

The finest constructional timber grows on the seaward slopes of the Coast Range and on Vancouver Island, where it is easily accessible from the sea. Lumbering takes place throughout the year, and most of the logs have to be moved some distance overland by tractor or light railway, or via 'skidways' as far as deep water, and are then rafted to the sawmills. The smaller and less valuable trees supply pulp and paper mills.

British Columbia provides 70 per cent of Canada's timber exports, most of which go to the United States. Large quantities of timber are also used in local woodworking factories which make veneers, plywood, furniture, doors and window frames. Power supplies are obtained from hydro-electric stations on the rivers and lakes.

As in eastern Canada, in the early days there was indiscriminate felling of the more valuable trees, but almost all the forest land is now owned by the provincial government (though it is managed by private companies) and cutting is accompanied by systematic re-

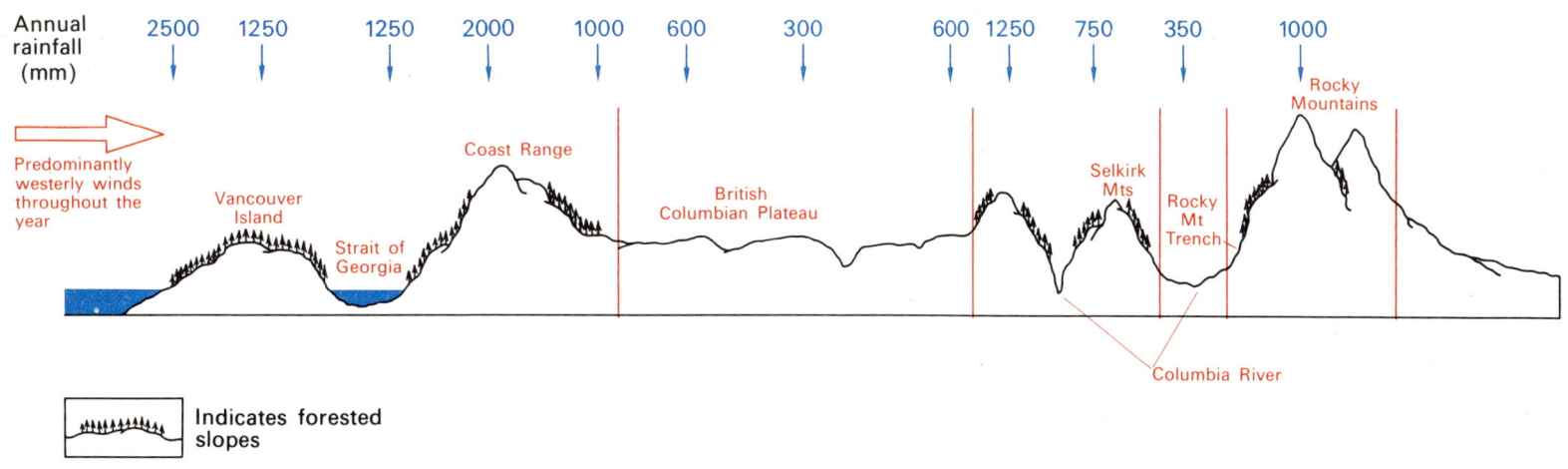

Annual rainfall (mm)

2500 1250 1250 2000 1000 600 300 600 1250 750 350 1000

Predominantly westerly winds throughout the year

Vancouver Island

Strait of Georgia

Coast Range

British Columbian Plateau

Selkirk Mts

Rocky Mt Trench

Rocky Mountains

Columbia River

Indicates forested slopes

FIG. 9.1 **A section from west to east across southern British Columbia**

Farming

planting. Efficient fire-fighting organisations are maintained, with look-out towers and spotter aircraft.

Because of the mountainous topography, only a very small proportion of the land can be used for farming, but in spite of this, agriculture occupies an important place in the province's economy. Mixed farming and dairying are carried on in southern coastal areas, especially on Vancouver Island and in the lower Fraser Valley, where the mild, moist conditions favour the growth of meadow grasses and fodder crops. Many farmers in these areas are also engaged in vegetable growing, and some in poultry and egg production. The considerable population of the towns provides a ready market for such produce.

In some of the southern valleys, including the Okanagan and Kootenay Valleys, where the summers are warm and sunny, there is large scale fruit growing. Melting snow in the mountains feeds the streams, which are used to irrigate orchards of apples, pears, plums, peaches and apricots, large quantities of which are canned for export. Because of the danger of late spring frosts, most of the orchards are planted on the valley sides, and the valley floors are used for vegetable growing and cattle pasture. In contrast to these irrigated valleys, most

of the drier interior is suitable only for rough grazing land for cattle and sheep.

Fishing

British Columbia's fisheries provide nearly 40 per cent of the total Canadian catch measured by value, though considerably less when measured by weight. The reason for this is the high value of the chief catch, salmon. These fish are hatched in the mountain streams, but spend most of their adult lives in the North Pacific Ocean before returning to the streams and lakes where they were born, to spawn and die. Large numbers are caught off the river mouths as they swim in, some in nets and some by line. Most of the catch is canned for export, the rest being frozen for sale in Canada and the United States. Thoughtless over-fishing in the past led to a serious fall in numbers, and the building of hydro-electric dams on the rivers discouraged spawning. Fishing is now restricted, and stairways have been provided at many of the dams. Special salmon hatcheries have been established from which the small fry are taken to the streams.

Next in importance to salmon comes halibut which, together with herring, tuna, mackerel and pilchard, are caught in deeper water by vessels operating from Prince Rupert, New Westminster, Victoria and

other ports. Most of the herring catch, which is much greater in
weight than the salmon catch, is processed into fish meal and oil.
Whaling and sealing were formerly important, but excessive killing
has greatly reduced the numbers of these animals.

Mineral Working

There is a great variety of mineral wealth in British Columbia. In the
nineteenth century the discovery of gold in the Fraser Valley and
Caribou Mountains brought thousands of new settlers to the region,
but today gold mining is much less important than the mining of zinc,
lead and silver. Eighty per cent of Canada's lead and half its zinc are
produced at the Sullivan Mine near Kimberley, west of the Rocky
Mountain Trench. Smelting is concentrated at Trail, which has the
largest lead and zinc smelter in the world and also treats copper ores
from the Coast Range, and silver, lead and zinc ores from the Yukon.
Formerly much damage was done to the surrounding vegetation by
emissions of sulphur dioxide from the smelter, but a new plant now
converts this into sulphuric acid and ammonium sulphate, the latter
being a valuable fertiliser.

Coal is mined near Nanaimo on the east side of Vancouver Island
and near Crow's Nest Pass, but although the coal is of good quality
and the seams thick, the output has remained small because of
competition from oil, natural gas and hydro-electricity. Oil is received
by pipeline from Alberta and natural gas from the Peace River
District.

Hydro-electric Power

There are abundant supplies of hydro-electricity in the province,
although only a small fraction of the power available has so far been
developed. Much of it is produced at small hydro-electric stations,
but three major schemes deserve special mention.

One of the largest is sited about 200 kilometres south-east of Prince
Rupert. Here the Nechako River has been dammed, forming a huge
lake from which water is led through a 16 kilometres tunnel to an
underground power station at Kemano. The electricity is supplied to a
large aluminium works at Kitimat, which lies on the Douglas Channel
and imports alumina directly from Jamaica and Guyana.

A second major project is the harnessing of the Peace River by the
construction of the great Portage Mountain Dam, behind which the
largest lake in British Columbia will form. When the scheme is
completed in the mid-70s it will be one of the most powerful hydro-
electric stations in the world.

The Photographic Survey Corporation Ltd, Toronto
**The Columbia Ice Fields and Rocky Mountains, along the border
between Alberta and British Columbia**

National Film Board of Canada
**The growing of apples and other fine quality fruit in South
Thompson valley near Kamloops**

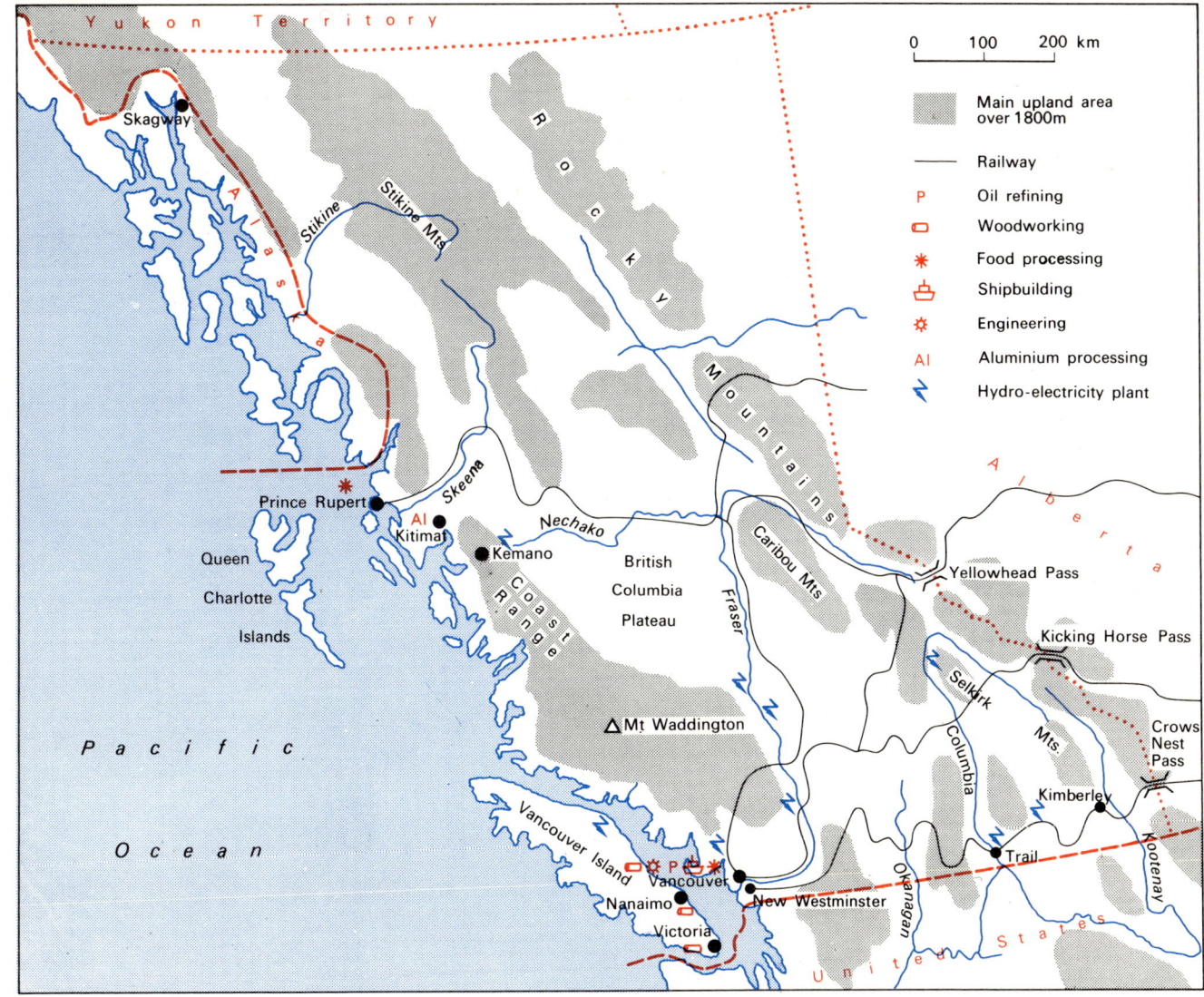

FIG. 9.2 **British Columbia**

A third scheme is the joint Canadian–United States Columbia River Project, an integrated plan agreed in 1961 but still in its early stages. The Columbia River is variable in volume and has caused serious flooding in the past, and the scheme is aimed at flood control as well as the supply of hydro-electric power and irrigation water. As its share of the plan, Canada will construct three huge dams:—Duncan Dam on the Duncan River, and Arrow Lake and Mica Dams on the Columbia River, as well as several smaller ones (see figure 19.3).

Population and Towns

According to the 1966 census, British Columbia had a population of 1 874 000, which is very small in view of the large size of the province and its wealth of natural resources. One of the chief handicaps to development is the great distance from the large markets of eastern Canada and the United States, and also the difficult communications through the Western Cordillera. Three-quarters of the population live within 160 kilometres of Vancouver, and most of the rest in the southern valleys and a number of isolated mineral-working towns.

Vancouver (892 000) is by far the largest city, and is situated on the southern side of the deep, sheltered Burrard Inlet, just north of the delta of the Fraser. It is overlooked by forested mountains, but the Fraser Valley affords easy access to the interior. It is the terminus of the Canadian Pacific Railway and of the southern branch of the Canadian National Railway and its main exports are wheat, timber, metals and canned fish. Much of its trade is with the Far East and Australasia, but some is with Europe via the Panama Canal. There are a number of coastal shipping services including one to the Alaskan port of Skagway, from which the Yukon can be reached by railway. Vancouver's principal industries are timber working, engineering, shipbuilding, oil refining and food processing. A short distance to the

The Photographic Survey Corporation Ltd, Toronto
Downtown Vancouver and its harbour (above)

National Film Board of Canada
Salmon seiner off coast of Vancouver Island (left)

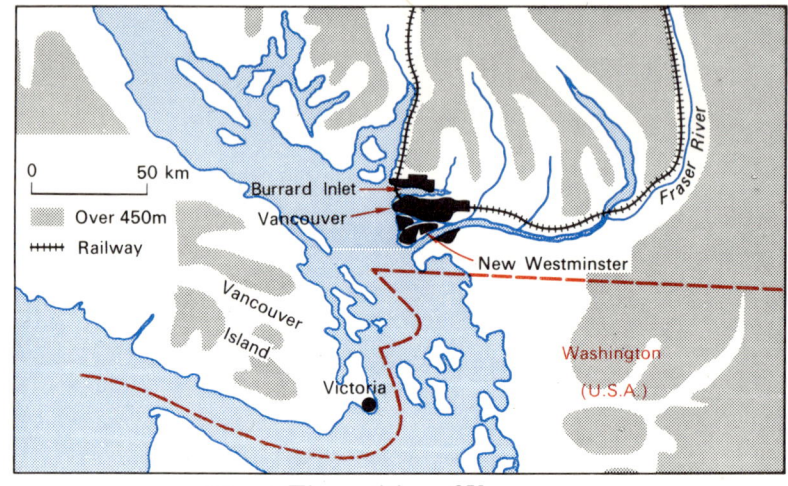

FIG. 9.3 **The position of Vancouver**

56

Alcan

The aluminium smelter at Kitimat

An artist's drawing of the Kitimat hydro-electric scheme, showing the tunnel and power station at Kemano

Draw a simple plan of the scheme, based on these photographs, and explain how favourable geographical conditions have been utilised to produce aluminium at Kitimat.

south, at the head of the Fraser River delta, is the smaller port of *New Westminster*.

Victoria (173 500) lies at the south-eastern end of Vancouver Island and is the capital of British Columbia. Its main function is as a government and administrative centre, but it is also a port of call for liners, a fishing port and naval base. *Prince Rupert* (15 000) is a port at the mouth of the Skeena River, and is the terminus of the northern branch of the Canadian National Railway. Fishing is carried on from the port and there is a salmon canning industry.

The Yukon

The Yukon Territory is a rugged plateau with many high mountains including Mount Logan (6050 metres) which is the highest peak in Canada. It has a very severe climate, with mean annual temperatures below freezing point, and a vegetation of scattered conifers or tundra.

In 1896 gold was discovered in the alluvial deposits of the Klondike tributary of the Yukon and a great gold rush began. By 1900 production was at its peak and Dawson City, the chief mining centre, had a population of 27 000. Today Dawson is a 'ghost town' and the whole of Yukon Territory has a population of only 15 000. Mining is still carried on, but silver now occupies first place, followed by lead, gold and zinc. Other activities are trapping and fur farming, with a little agriculture in the vicinity of the mining camps.

The Yukon River is the chief highway in summer, when steamer services run from Dawson to Whitehorse, the administrative centre. Whitehorse is linked by narrow gauge railway with Skagway in Alaska, and with the Prairies by the Alaska Highway, which was built for military purposes during the war and is now used for bringing in supplies.

Exercises

Answer in note form :
1. Which are the most important farming areas in British Columbia, and what kinds of farming are practised?
2. Compare the forest industries of British Columbia with those of Newfoundland.
3. Describe the site and position of Vancouver.
4. Write an account of the fisheries of the Canadian Pacific coast.
Essay :
  What are the principal natural resources of British Columbia and to what extent do they await development?

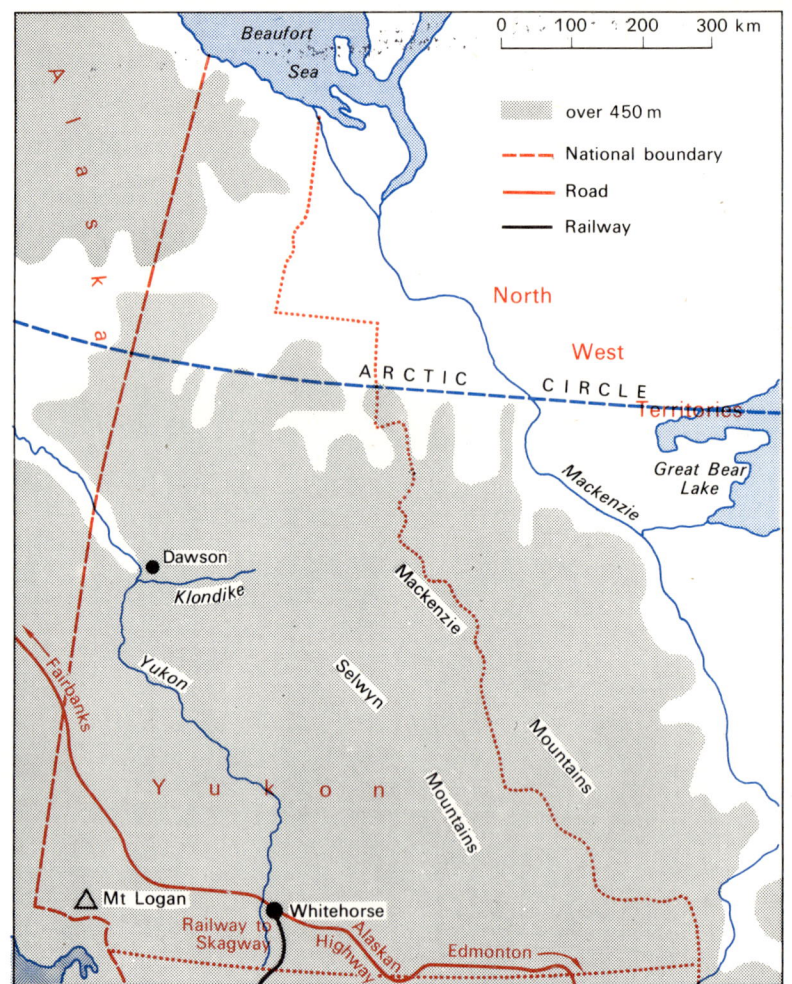

FIG. 9.4 **The Yukon**

10: THE UNITED STATES – GENERAL ECONOMIC SURVEY

The first European settlements in what is now the United States date from the beginning of the seventeenth century when small colonies were established round the shores of Chesapeake Bay and farther north in New England. During that century a steady stream of immigrants came from Europe and settlements gradually spread down the coast, although movement far inland was very limited. This was partly because of the rugged Appalachian ranges and the presence of hostile Indian tribes, and also because there was still plenty of opportunity on the Atlantic seaboard. Meanwhile the French, who had built up a settlement at New Orleans, laid claim to a vast stretch of territory including most of the Mississippi Basin.

The United States originated from the thirteen Atlantic states which were responsible for the Declaration of Independence in 1776. In 1803 the boundary was greatly extended by the purchase of much of the Mississippi basin from the French. Texas became part of the Union in 1845 and California in 1847, both having previously been Spanish and then Mexican possessions. Thus by the mid-nineteenth century the map of the United States began to look as it does today. In 1867 Alaska was acquired from Russia and in 1959 it became the forty-ninth state, followed a year later by Hawaii.

The rapid population growth of the United States owes much to immigration. The original immigrants included many adventurers and people escaping from religious persecution, as well as practical businessmen and farmers. Soon the expansion of the slave trade brought in African slaves to work the plantations on the southern Atlantic coast. The stream of immigrants in more recent years has been irregular, coinciding with periods of economic depression or war: for example many Irish came over in the nineteenth century as a result of famine and unrest in Ireland, and many Russians arrived after the Russian Revolution of 1917. Already during the present century alone, 25 million immigrants have entered the United States from Europe. At the same time large numbers from Central America, especially Puerto Rico, have crowded into the cities of the north-east, and many Chinese and Japanese have crossed the Pacific to settle in California.

The United States has an area of 9 192 000 square kilometres and a population of about 200 million, which gives an average density of about 22 persons to the square kilometre. The population is drawn from all the countries of Europe, and also includes some 22 million negroes, 500 000 Red Indians, 150 000 Japanese and 100 000 Chinese. It is a land rich in natural resources, with a high proportion of fertile farmland, great mineral wealth, hydro-electric reserves and forests, and its people have shown remarkable energy in developing these resources. America is outstanding in the field of technology and makes use of the most advanced production methods in its industries. As a result, both in industrial and farm output, it leads the world.

The distribution of population is shown in Figure 10.1 and the following points should be noted:

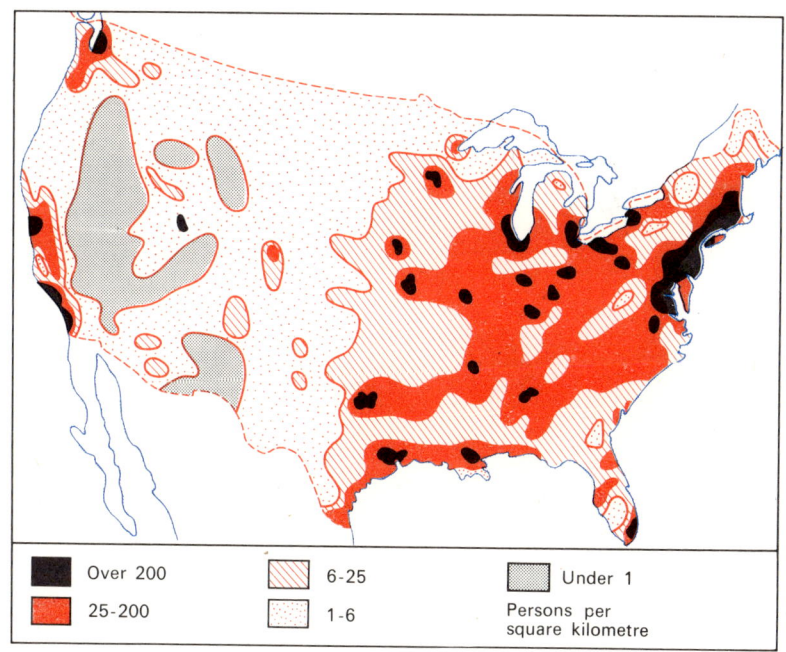

■	Over 200	▨	6-25	▨	Under 1
■	25-200	⠂	1-6		Persons per square kilometre

FIG. 10.1 **The United States – distribution of population**

(1) The great concentration of population in the urban belts of the north-east. One belt extends from Boston to Washington along the Atlantic coast, and the other from Chicago to Pittsburgh and the southern shores of Lake Erie.

(2) Areas which have experienced great expansion during the present century, notably the Pacific coast and Florida, as well as the western coastlands of the Gulf of Mexico and the Piedmont Plateau immediately east of the Appalachians.

(3) A number of great urban concentrations in the Mississippi Basin, with a well developed pattern of rural settlement.

(4) An absence of settlement in the western mountain zone except for isolated mineral workings and irrigation areas.

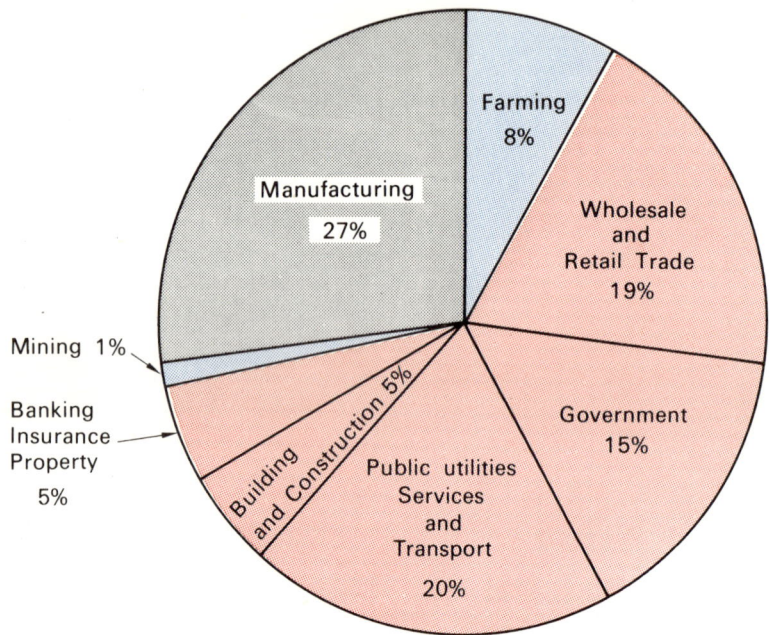

FIG. 10.2 **The United States – employment (percentage of working population)**

Farming

Only 8 per cent of America's working population is employed on the land. Farm mechanisation has led to a steady fall in the number of workers, and yet agricultural output continues to increase with the

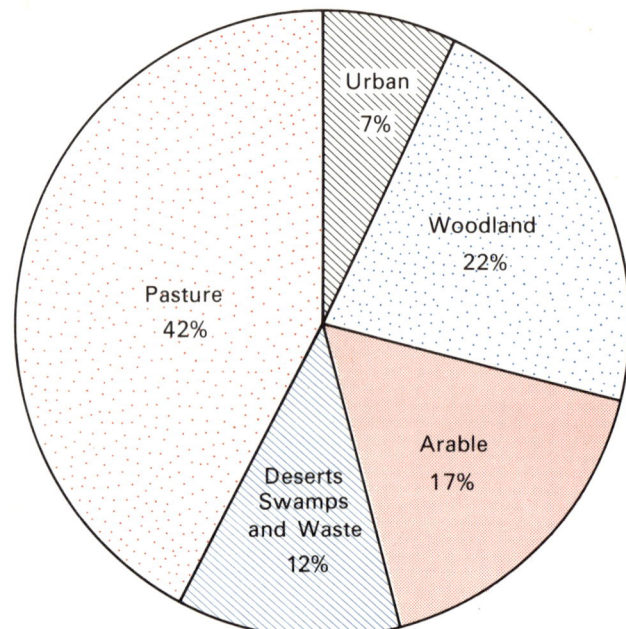

FIG. 10.3 **The United States – land use**

use of more intensive methods. The United States leads the world in the production of maize, oats, cotton, tobacco, citrus fruits and livestock. The principal types of farming are shown in Figure 10.4, but it must be realised that these represent only the main activities; for example, throughout the maize belt some wheat is grown as well as oats and barley, and in some places soya beans and sorghums occupy as much land as maize.

Crop Production

Spring wheat growing is concentrated in North and South Dakota, whilst winter wheat, which requires less severe conditions, is grown farther south, particularly in Kansas, but extending right across the maize belt. Maize itself is grown widely throughout the United States, even in areas in the north-east where the cob rarely matures but where the stem and leaves provide a nutritious fodder especially for silage. Cotton production is located mainly on the Mississippi flood plain, in north-west Texas and, under irrigation, in Arizona and California.

More intensive forms of farming are used in the cultivation of sugar beet, which is grown mainly in narrow irrigated belts in the Western Cordillera and central California, and sugar cane which is produced in Louisiana, Florida and Hawaii. Market gardening (known as truck farming) is concentrated particularly (a) near large centres of population, especially in the north-east, (b) in irrigated areas of the High Plains, Western Cordillera and California, and (c) on recently reclaimed land in Florida. Although fruit farming occupies only a small proportion of land, production is enormous. The United States produces one-third of the world's citrus fruit, mainly in California and Florida, whilst apples are grown in Washington state, in the northern Appalachian valleys, in Michigan and along the shores of Lake Erie. Tobacco is another crop which occupies only a small acreage but is of great value, being grown mainly in North and South Carolina, Virginia and Kentucky.

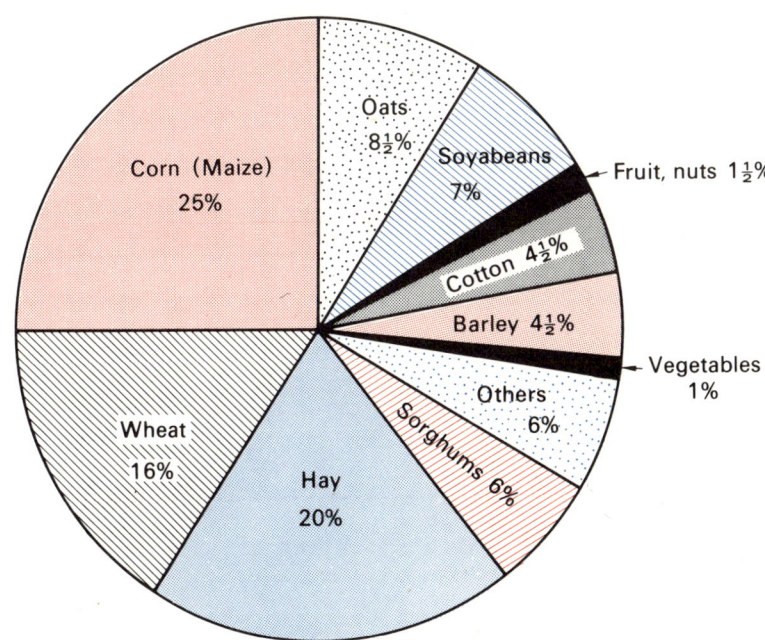

FIG. 10.5 **The uses made of cropland in the United States**

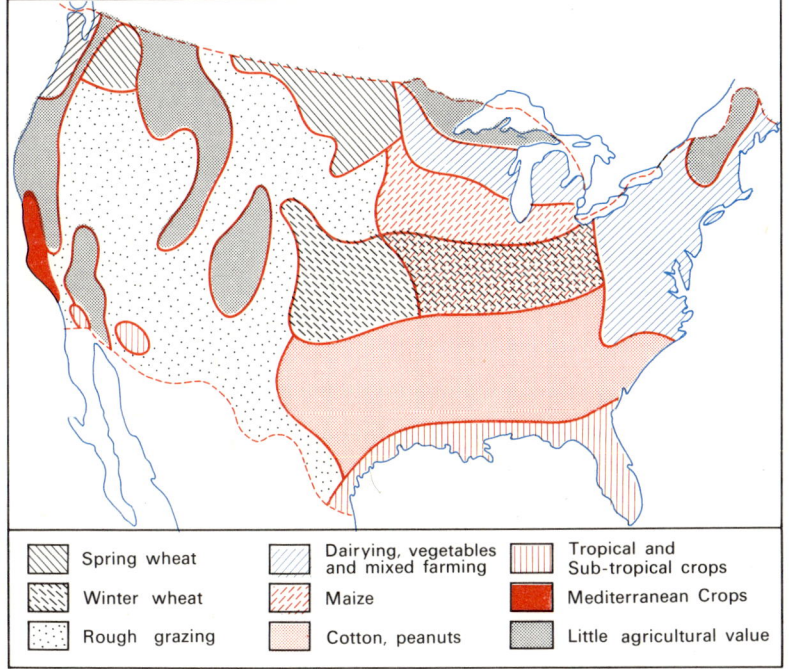

FIG. 10.4 **Main agricultural regions of the United States**

Spring wheat

Winter wheat

Rough grazing

Dairying, vegetables and mixed farming

Maize

Cotton, peanuts

Tropical and Sub-tropical crops

Mediterranean Crops

Little agricultural value

Livestock

Most of the sheep are relegated to the poorer grazing lands especially in the High Plains and Western Cordillera, with the greatest concentration on the Edwards Plateau in Texas. On the other hand, cattle have a very widespread distribution, dairy cattle being especially numerous near the great urban belts of the north-east where the rainfall is adequate to maintain a rich pasture for most of the year. Minnesota, Wisconsin and Michigan are the leading dairying states. Both pigs and beef cattle are concentrated in the maize belt although beef cattle have also increased in number in the southern states, particularly Florida and Georgia. Many beef cattle are reared on poor pastures in the Western Cordillera and are usually sent to the maize belt for fattening.

Lumbering

The United States is second after the U.S.S.R. for the production of softwood and it is also one of the leading producers of hardwood, but even so large quantities of timber and wood pulp have to be imported

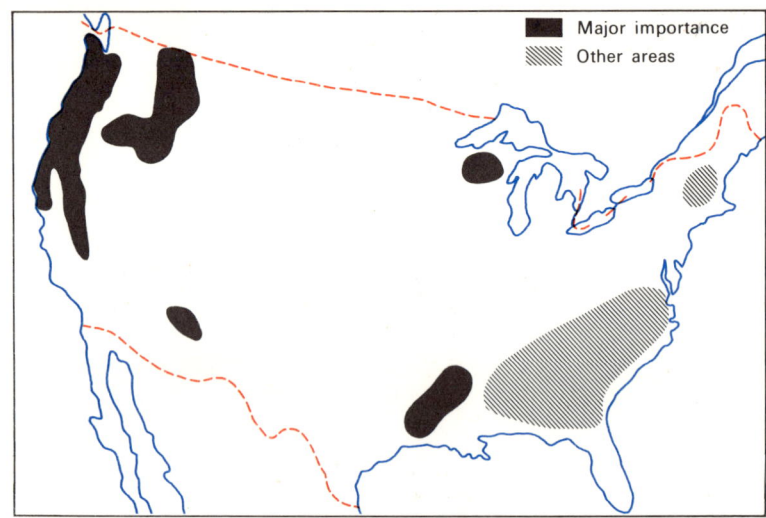

FIG. 10.6 **Main lumbering areas**

to meet the enormous home demand. The main producing areas are:

(1) The North-West including Washington, Oregon and north California. This is by far the most important region today for timber production, and most of the wood is used for constructional purposes. The industry is organised on a large scale with many vast milling concerns. The main trees are Douglas Fir, redwood, hemlock, cedar, spruce and pine.

(2) The next most important area is in the South-East in a belt extending from Louisiana to Virginia. Here mainly softwood (cypress, southern pine), but some hardwood as well (oak, hickory, walnut, cherry) is cut. The lumbering operations and mills are numerous but small scale, and much of the timber is pulped for making into paper.

(3) In the north-east, particularly the New England states. This was the first part of the country to have a lumbering industry and much of the good timber, which included pine, spruce, hemlock, maple, oak and hickory, has been cut. The main use of the wood is now in pulp and paper mills which are increasing their import of timber to survive.

Fishing

America is one of the major fishing nations of the world. There has been some cutback in the industry due partly to overfishing and also to pollution which has particularly affected the inshore fisheries. There is some fishing off all the shores of the country. From the New England ports the chief fish caught are haddock and herring, and farther south along the mid-Atlantic coast the most important are menhaden and shellfish. The Gulf of Mexico yields large quantities of menhaden and shrimps, whilst on the Pacific coast the Californian fishing fleets land tuna, mackerel and sardines. In Washington and Oregon the main catch is salmon, followed by halibut.

Minerals and Power Supplies

So great and varied are the mineral deposits and energy resources that this general account can attempt only to point out the outstanding features. At present the United States is the most powerful industrial nation in the world and this is largely the result of its enormous supplies of three basic commodities – coal, iron ore and mineral oil.

About a quarter of the world's annual production of coal is mined in the United States, the output being in the region of 450 million

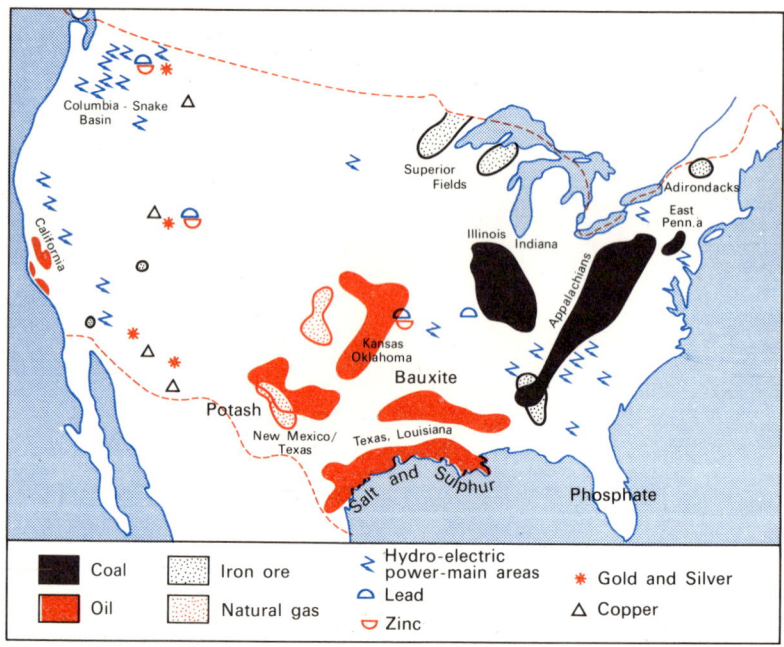

FIG. 10.7 **The United States – minerals and power resources**

tonnes. The Appalachian field is by far the most important and the seams here are thick, uniform and near the surface, so that mining methods can be highly mechanised. As a result, production per man-shift is three times greater than in Western Europe. Output, now declining in most West European countries, continues to rise here, much of the increase being accounted for by the demands of thermal electricity power stations.

In iron ore production the United States is second only to the U.S.S.R., but in spite of this, and the use of large amounts of scrap, imports of iron ore are increasing year by year. The greatest output is from the Lake Superior ores which are obtained by open-cast working and have yielded far more ore than any other deposit in the world. Inevitably, the better quality ores are being rapidly exhausted which accounts for the growing imports of Canadian and Venezuelan ores. Other important sources of iron ore are in the northern and southern Appalachians and Western Cordillera.

America has dominated petroleum production throughout this century, and produces about one-quarter of the world's mineral oil. The present output is more than 400 million tonnes per annum, and it is still rising. The boom in the industry began with the early days of the motor car and continued with the switch from coal to oil as a fuel on the railways and for ships. Today it is much used for heating and as a raw material in the petro-chemical industry. The country's proportion of the world's natural gas is even greater than its oil, amounting to some 60 per cent. The gas, like the oil, is distributed by pipeline and is also being increasingly used for heating and as an alternative raw material in the petro-chemical industry.

Figure 10.7 shows the distribution of the major hydro-electric stations, and it will be seen that most developments have taken place in limited areas remote from alternative sources of energy. The most important area for hydro-electricity is the River Columbia basin, followed by the Tennessee Valley.

The United States is the leading producer in the world of copper, phosphate, potash, sulphur and salt, and a major producer of lead, zinc, bauxite, gold and uranium. The distribution of the main deposits of these minerals is shown in Figure 10.7.

Manufacturing

Before the last war it was true to say that the bulk of the manufacturing industries of the United States were concentrated in the north-east in a triangle bounded by Chicago, Washington and Boston, but today

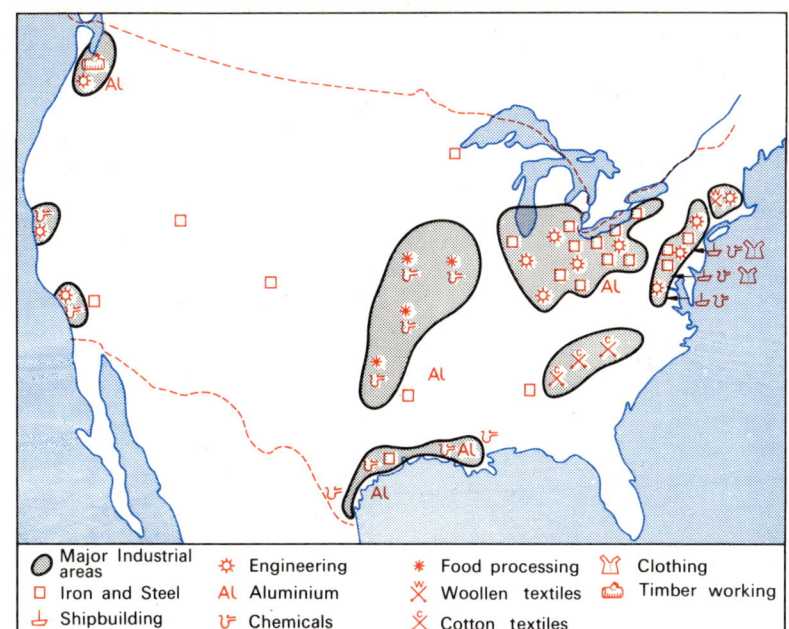

FIG. 10.8 **The manufacturing regions of the United States**

manufacturing has become far more dispersed. Nevertheless, the iron and steel industry is still largely located in this industrial triangle, with Chicago and the Pittsburgh–Youngstown districts as the leading producing areas, lying where local coal and Lake Superior iron ore can easily be brought together. The expansion of the industry on the mid-Atlantic coast reflects the increased dependence on imported ores as well as the enormous local demand for steel. Other more scattered steel industries have grown up because of local raw materials sometimes coupled with expanding markets, as at Fontana near Los Angeles. Total steel production reached 119 million tonnes in 1968.

The distribution of engineering industries is far more widespread. Besides heavy engineering firms which benefit from being near to the steel producing sites, many factories, particularly those concerned with machinery, aircraft and electrical goods, have developed in areas of recent population expansion notably in California, Florida and some of the boom towns of the mid-west.

The textile industry was formerly concentrated in New England, but local supplies of raw cotton, cheaper labour and modern machinery led to the expansion of the cotton industry on the Piedmont Plateau of North and South Carolina, where synthetic fibres and woollen goods are now also produced. New England remains important for woollens whilst the mid-Atlantic coast region is concerned particularly with man-made fibres, showing the influence of the local petrochemical industry and the many clothing factories.

The chemical industry is widely dispersed throughout the eastern half of the country but certain location factors are evident. Oil refining has encouraged petro-chemical industries on the Gulf coast as well as at Los Angeles, whilst local phosphate deposits provide raw material for central Florida's chemical industry. Chemical manufacture is also important in the New York–Philadelphia area which benefits from a coastal location for the import of bulky raw materials including oil, and in West Virginia where the industry has been influenced by local coal.

Foreign Trade

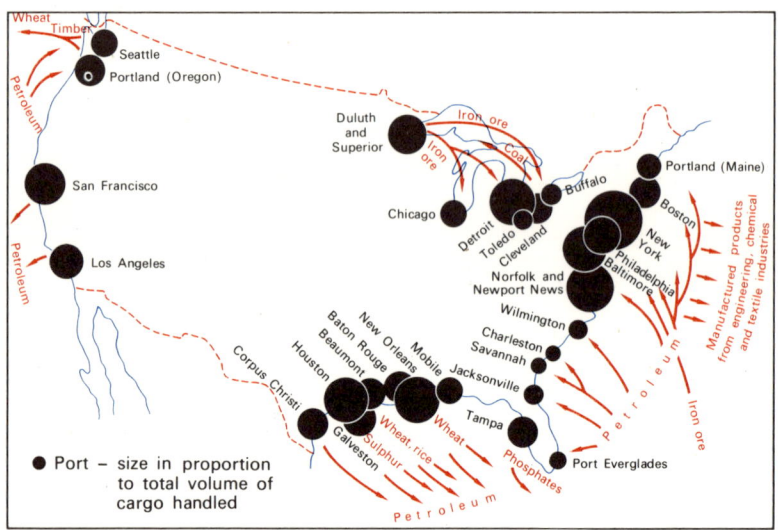

FIG. 10.9 **Major ports and trade movements**

Until the present century the United States was mainly an exporter of food and raw materials and an importer of manufactured goods, chiefly from Europe, but the rapid growth of American industry

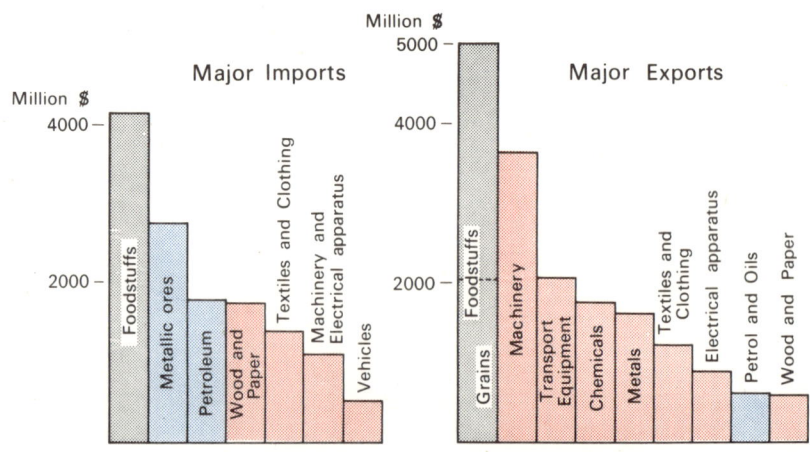

FIG. 10.10 **The United States – foreign trade**

during the last eighty years has completely altered the character of her overseas trade. Today manufactured goods figure prominently among the exports alongside certain farm products such as cotton, grain and tobacco. A considerable proportion of the imports consists of raw materials to supply the growing needs of American industry, including oil from Venezuela, Mexico and the Middle East, wood pulp, timber and mineral ores from Canada, wool from Australia and New Zealand, copper from Chile, bauxite from the Guianas and iron ore from Venezuela and Brazil. Other imports include tropical foodstuffs and some specialised manufactured goods.

Exercises

Answer in note form:

1. What are the main factors which have influenced the distribution of (a) the iron and steel industry and (b) the chemical industry in the United States?
2. Describe (a) the distribution of market gardening and (b) the main workable timber reserves in the United States.
3. What are the chief features of America's overseas trade?

Essay:

Analyse the distribution of population in the United States and explain why there is such a concentration in the north-east.

11: NEW ENGLAND

This small north-east corner of the United States has played an important part in the history and early development of the country, for it was here that the Pilgrim Fathers landed in 1620. The early settlers had to contend with a harsh climate, but they could readily obtain timber, fish and furs. Agriculture was possible on the limited

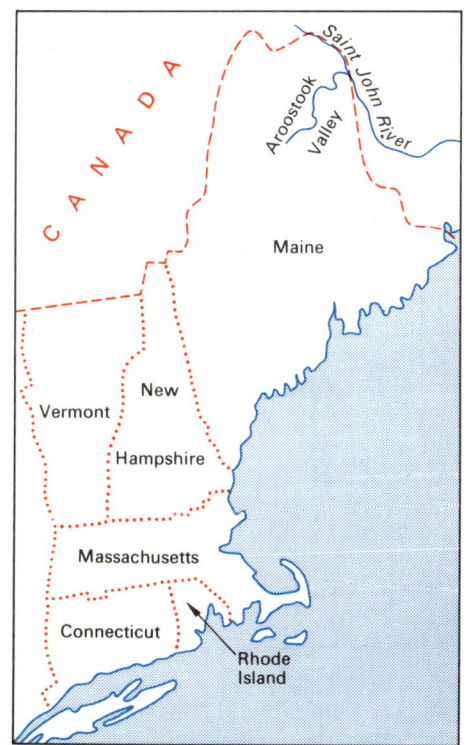

FIG. 11.1 **The States of New England**

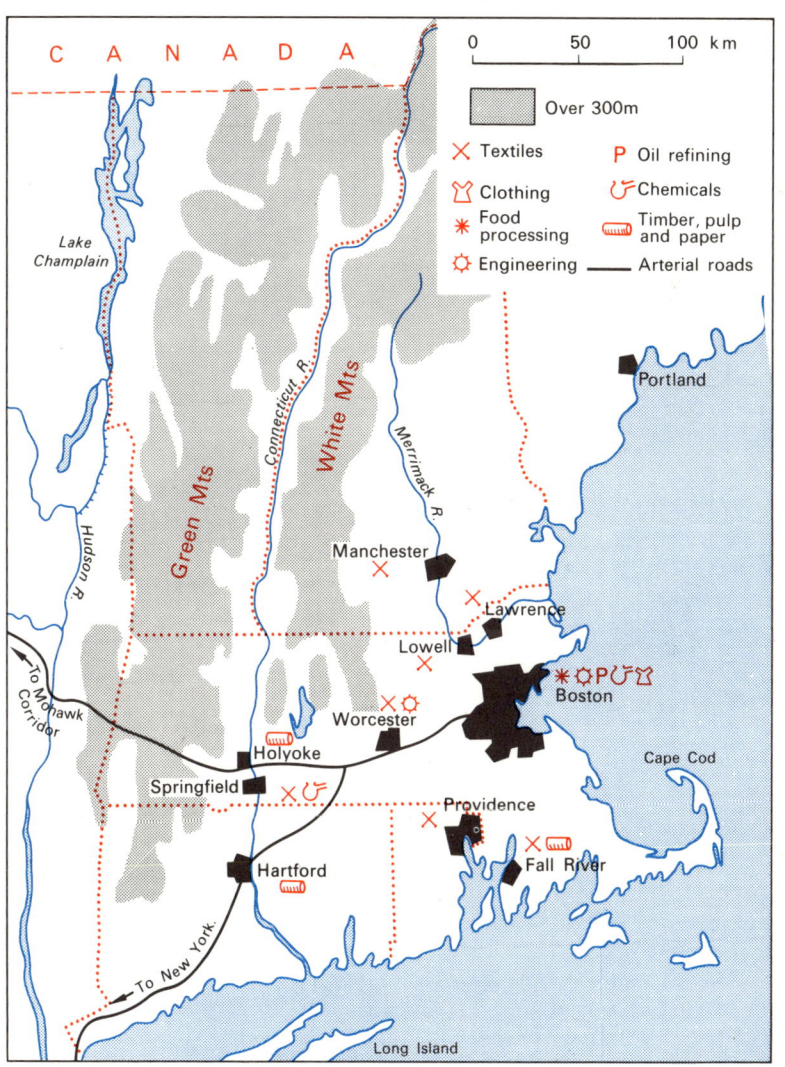

FIG. 11.2 **Southern New England**

U.S. Information Service

The White Mountains, New Hampshire
What aspects of the geography of the interior of New England does this photograph illustrate?

The interior uplands are made up of ancient, resistant rocks such as granite, slate and marble, and are a continuation of the Appalachian mountain system, but the main trend here is from north to south. The Green and White Mountains rise to between 1000 and 1500 metres above sea level. The scenery is often spectacular with precipitous slopes and gorges, and this has become a popular recreational area serving the large populations of the surrounding lowlands. The uplands of Maine are lower but equally rugged.

The climate of New England is typical of the cool temperate eastern margins, with long, severe winters, cool summers and fairly heavy precipitation ranging from 900 to 1400 millimetres per annum. As an example, Boston has a mean January temperature of −3°C, a July mean of 22°C, and an annual precipitation of 1133 millimetres. Temperatures are lower in the interior uplands where heavy winter snowfall coupled with blustery winds give severe conditions.

Originally New England was thickly forested with maple, oak, hickory and other deciduous trees on the lowlands, and conifers, including pine, spruce and hemlock, higher up. Most of the lowlands have now been cleared although there are large tracts of woodland on the uplands, particularly in the north, where over three-quarters of Maine and New Hampshire are still forested.

Lumbering

Lumbering is declining, as much of the more accessible and finer quality timber has been cut. The pulp and paper industry remains important, although timber has to be imported into the region to supplement local supplies. There is increasing competition from the western and southern regions of the country where there are more plentiful supplies of timber and the mills have more up-to-date equipment. The main centres are near to hydro-electric power sites and include Holyoke and Fall River.

Farming

Farming has never been a leading occupation but New Englanders have developed specialised activities on the limited amounts of fertile land to supply the large town populations of the north-eastern United States. The Connecticut Valley is one of the most important farming areas, the main activities being dairying, poultry farming and the growing of vegetables. A long established specialisation here is tobacco production; yields are heavy and many of the plants are grown under cheese cloth to give a strong leaf suitable for use in cigars. The Cape

amounts of fertile land, but prosperity required effort and the New Englanders have been noted for hard work. On the other hand, there has been a conservative attitude in the face of new ideas which, together with the limited industrial resources, especially the lack of fuel, brought about the relative decline of the region during the early twentieth century.

The coastal areas of New England consist of low lying, undulating country. Glaciation has left irregular deposits of gravel and clay, and poor drainage has resulted in stretches of coastal swamps and many lakes. In the north the coast of Maine is more rugged and broken.

U.S. Information Service

Potato harvesting in the Aroostook Valley, Maine
This is a major producing area for this crop and the centre for seed potatoes.

Cod area is noted in particular for cranberries, and the Aroostook Valley in north-east Maine for potatoes, whilst within the upland valleys there is a considerable amount of dairying.

Fishing

Fishing has been important since the time of the first settlers although there has been some decline in recent years. The region has the advantage of a highly indented coast providing good harbours, rich fishing grounds both inshore and on the 'Banks', and large local markets. Boston and Portland are the chief fishing ports but there are many smaller ones, and the catches include haddock, cod, herring and shellfish, especially lobsters. Fishing has led to numerous processing activities such as freezing, curing, and the manufacture of fish meal and fertilisers.

Manufacturing and Towns

In spite of limited resources, manufacturing is the basis of New England's economy. The reasons for its importance are partly historical. The first factories were textile mills employing water power.

With the advent of steam power it proved cheaper for the owners to enlarge these factories and bring in coal for raising steam, than to move to the coalfields. New England also had the advantage of a highly skilled labour force and a reputation for quality products. Nevertheless industrial growth was checked at the end of the nineteenth century and during the present century there has been a marked change in its character.

Now the southern states with modern machinery and cheaper labour have taken the lead in cotton manufacturing from New England, but the latter remains a major woollen manufacturing area and the production of man-made fibres has expanded in recent years. The manufacture of clothing, especially high quality articles, is well represented in most of the large towns of the region, and another long established industry is the production of boots and shoes, which accounts for one-third of American production. Manchester, Providence and Fall River are leading textile centres.

U.S. Information Service

Gloucester, Massachusetts
This is typical of many small fishing ports which occupy sheltered inlets along the New England coast.

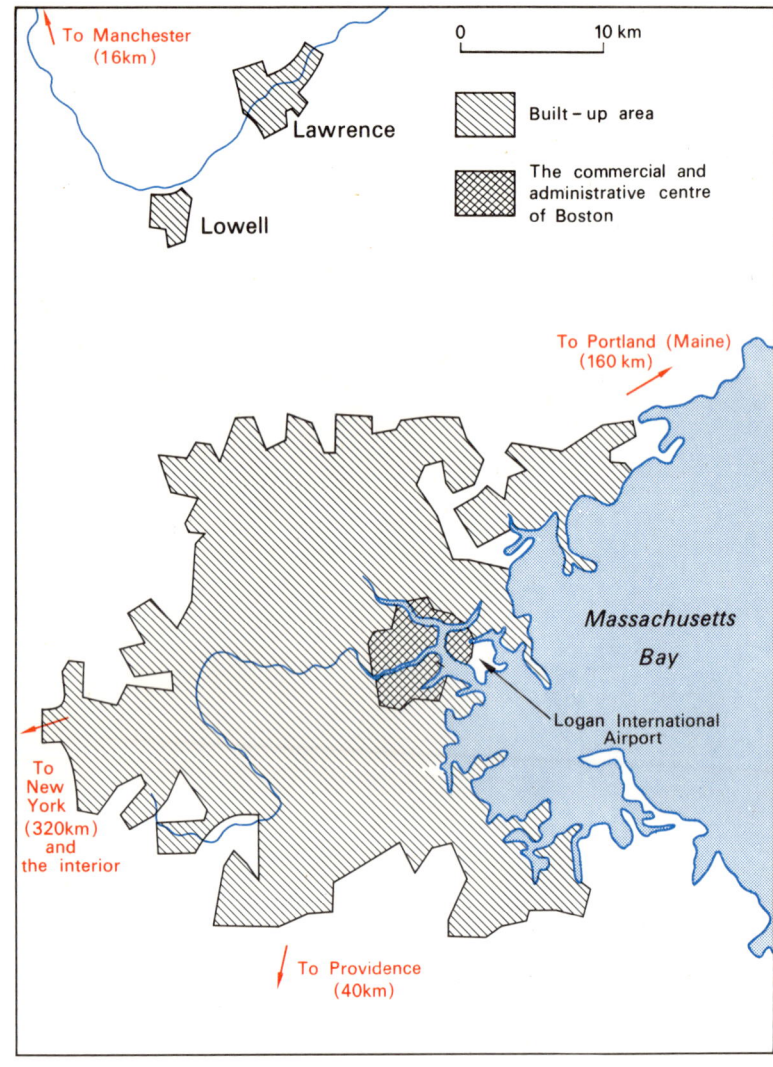

To Manchester (16km)

Lawrence

Lowell

0   10 km

⬛ Built-up area

⬛ The commercial and administrative centre of Boston

To Portland (Maine) (160 km)

Massachusetts Bay

Logan International Airport

To New York (320km) and the interior

To Providence (40km)

FIG. 11.3 **Boston**

Today the biggest employer of labour is engineering, with an emphasis on precision work including the manufacture of machine tools, electronic equipment and computers. Expansion in these fields has been partly influenced by the research work carried on at the local universities of Harvard and Yale, and at the Massachusetts Institute of Technology, but engineering developed originally to supply the demand for textile and leather-working machinery. The chemical industry is small but rapidly expanding and a third of the nation's plastics output comes from this region.

The population of New England is most concentrated in Massachusetts, Connecticut and Rhode Island, but although the total population is considerable there is only one large city, *Boston*. The city itself has a population of three-quarters of a million, but if the surrounding urban district is included there are two and a half million. Boston grew up round the mouths of two small rivers flowing into Boston Bay, where there was deep, sheltered water close inshore as well as an extensive frontage for docks. It became the focus of routes in New England although inland links are naturally poor because of the north-south trend of the mountain ranges. Tunnels and canals have now been constructed to overcome these physical barriers. The town is an important commercial centre and the principal woollen and leather market of the United States. Its manufacturing industries include clothing and leather as well as various port industries.

Portland is the principal port of Maine and is well placed to serve the St Lawrence Valley. It provides a winter outlet for Canadian wheat and timber, and considerable quantities of Venezuelan oil are piped from the port to Montreal.

Exercises

Answer in note form:

1. Describe the chief features of the relief and natural vegetation of New England.
2. Make a table which summarises the main manufacturing industries of New England and where they are found.
3. Draw an annotated sketch map to illustrate the site, position and function of Boston.

Essay:

How has the physical geography of New England influenced the distribution and occupations of its inhabitants?

12: THE MID-ATLANTIC COASTAL PLAIN

This region, whose limits are shown in Figure 12.1, is one of the most densely populated in the world and contains some 20 per cent of the population of the United States. It has been described as the city belt of America and is a great industrial region, producing goods of every variety. Many of the people are recent immigrants and most of the rest are descended from immigrants who have come across the Atlantic in the twentieth century.

The relief of the region consists of undulating lowlands crossed by broad, shallow valleys, stretching from the piedmont zone of the Appalachians to the Atlantic coast. The rivers have brought down large quantities of silt and gravel from the uplands, whilst along the coast are extensive sand dunes and marshland. Glacial deposits are also present – for example, a pair of terminal moraines diverge from the eastern end of Long Island with an outwash plain of finer glacial material between them. The post-glacial rise in sea level is responsible for the highly indented coastline, including the large rias of Delaware and Chesapeake Bays.

The climate is fairly extreme with cold winters, during which heavy falls of snow occur, and warm, humid summers. To quote an example, New York has a mean January temperature of 0°C, a mean July temperature of 23°C, and an average annual precipitation of 1080 millimetres.

Farming

Although under 5 per cent of the working population of the region are engaged in farming, agriculture is of considerable importance to the economy. The vast local market has greatly influenced the character of the farming which is particularly concerned with perishable products including fresh fruits and a wide variety of vegetables. This type of farming, which in Britain is known as market gardening, is called truck farming in America because most farmers use motor trucks to carry the produce to market. The mid-Atlantic coastal plain

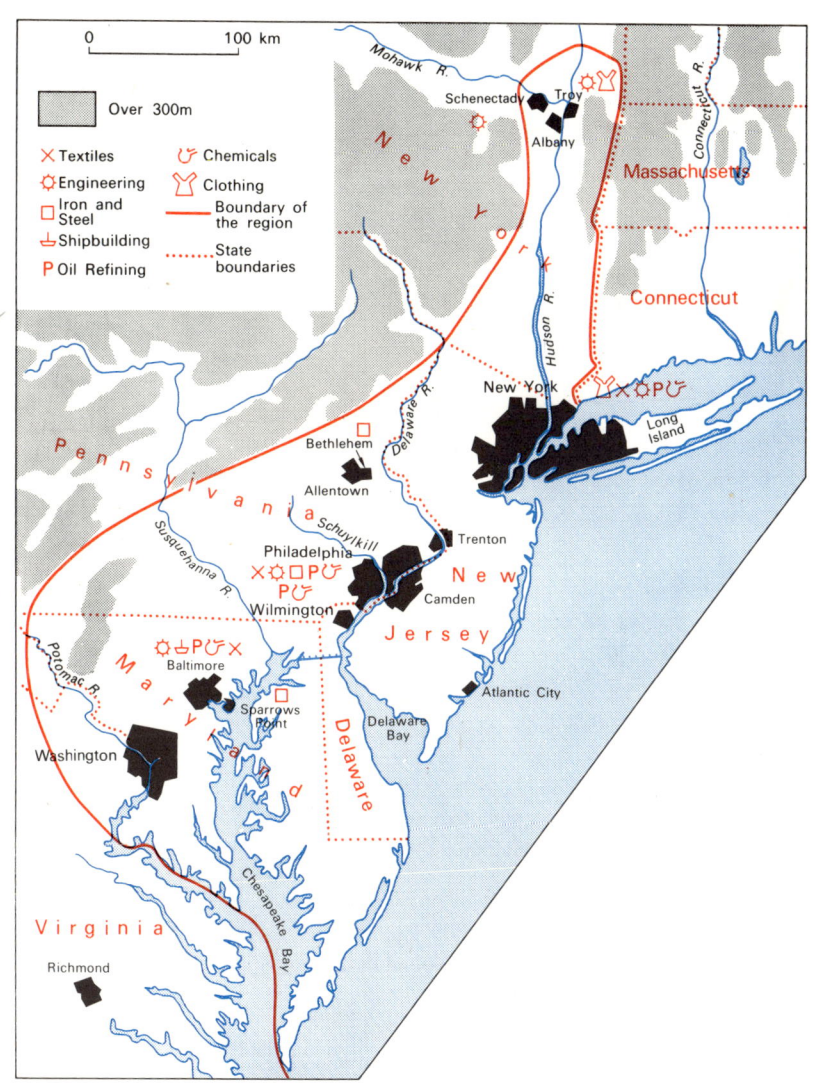

FIG. 12.1 **The Mid-Atlantic Coastal Plain**

U.S. Information Service
A dairy farm typical of many which supply the nearby urban areas of the north east with fresh milk. This farm is in Maryland.

has natural advantages for this specialisation. Over large parts of the region the soils are light, sandy loams which are easily worked and warm up quickly in spring, whilst the climate near the coast is milder than in the interior. Some of the main concentrations of truck farming are round the shores of Chesapeake Bay. The production of tomatoes is particularly important and the crop is so large that much of it is now canned.

The moist climate encourages the growth of rich grassland which has led to dairying for the supply of fresh milk to the nearby towns, and in recent years there has been a big increase in the production of broiler chickens, especially in the Delaware Peninsula. Along the interior edge of the region the Piedmont zone also has productive soils and here tobacco, dairying and fruit are all important in addition to more general forms of farming.

Fishing and Holiday Resorts

The shallow coastal inlets provide ideal conditions for shell fish, but overfishing as well as pollution have caused some decline in the fishing industry. However, oysters, clams, crabs and lobsters are still caught in large quantities, particularly in Delaware Bay and along the shores of Long Island.

The holiday industry is well developed and there are lines of resorts on the coast of New Jersey and Long Island. These provide an escape from the heat of the cities in the summer, offering sand, sea breezes and sunshine. One of the biggest, Atlantic City, has a permanent population of only 6000, but receives more than 10 million visitors each year, most of them on day outings.

Manufacturing and Towns

This is one of the most important industrial regions in the United States and the products are of great variety. In particular, it dominates the American clothing industry, whilst all types of engineering are represented, as well as the manufacture of chemicals including man-made fibres. The region gained an early lead in industrial development, for (a) it was the area to which most immigrants first came and it was consequently here that they first looked for employment; (b) the good natural harbours facilitated the import of raw materials; and (c) there were good communications with the interior and the Appalachian coalfield, especially via the Hudson and Mohawk valleys.

New York is one of the world's largest cities and has a population of about 8 million which includes over one million negroes, one million Italians, nearly a million Russians, and about half a million each of Irish, Poles and Puerto Ricans as well as large numbers of other distinct communities.

The city centre is on Manhattan Island, where the restricted space made it necessary to build upwards, hence the many skyscrapers. The city has also spread to the mainland as well as to Staten Island and Long Island. Thus there is a complex system of sheltered harbours and waterways providing many miles of water frontage.

New York lies at the southern end of the Hudson–Mohawk corridor, giving easy access to the Great Lakes, St Lawrence Valley and interior plains. Ocean-going vessels can reach Troy, 240 kilometres up the Hudson River, and from there the New York State Barge Canal leads

Manhattan Island, New York
Hudson River is on the left and East River on the right.

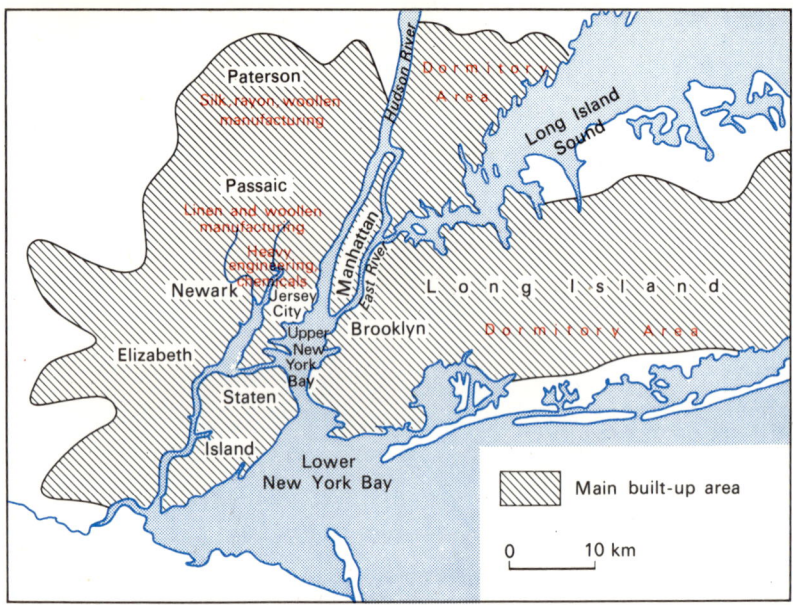

Paterson
Silk, rayon, woollen
manufacturing

Dormitory
Area

Hudson River

Long Island
Sound

Passaic
Linen and woollen
manufacturing

Heavy
engineering,
chemicals

Manhattan

East River

L o n g   I s l a n d

Newark
Jersey
City
Upper
New
York
Bay
Brooklyn
Dormitory Area

Elizabeth

Staten

Island

Lower
New York Bay

⧄ Main built-up area

0   10 km

FIG. 12.2 **The New York Conurbation**

to Buffalo and the Great Lakes. Railways and express roads follow the same route. Because of these advantages, New York is America's leading port. Its hinterland extends over much of the Mid-West and into Canada, and the deepening of the lower Hudson during the Ice Age has provided deep water close to the shore, so that no docks are needed. In addition, ships do not have to wait for high tide since the tidal range is only 1.25 metres. The most important wharves are on the Hudson and East Rivers, and a portion of Staten Island is a free trade area.

Because of its position at the crossroads of world trade routes and overland routes, New York has become the country's financial and commercial capital. Banking and insurance are outstanding activities and Wall Street, the nation's stock market, is the focus of much American business.

New York is the largest manufacturing centre in the United States. Its industries benefit from the excellent port facilities and good communications. The city itself is a great consumer market and the varied backgrounds and skills of its inhabitants are reflected in the great variety of activities. The most surprising feature is the very

small size of most factories, the average number of workers per factory being about twenty. The clothing industry employs more workers than any other. It produces three-quarters of all the women's clothing made in the U.S.A., and over one-third of the men's clothing. The industry is concentrated in a small part of the city and consists of thousands of enterprises, many of which occupy only a single large room or loft.

Other leading industries are the printing and publishing of books, magazines and newspapers, engineering, shipbuilding, food processing and the production of a wide range of consumer goods. The outer suburbs have developed more specialised industries, for example Paterson is noted for silk and rayon, Passaic for linen and woollen goods, and Jersey City for heavy engineering. The petro-chemical industry has developed rapidly in the Jersey City–New York area, encouraged by the oil refineries which have been established on reclaimed marshland.

Sprawling out from New York are belts and pockets of settlement across the whole of the region. Towards Philadelphia is a line of cities including New Jersey, Newark, Elizabeth, Trenton and Camden. This belt of residential and industrial development is an unattractive muddle of new and decaying property. A further line of settlement is along the Hudson River. The lower Hudson Valley has suburban areas interspersed with a great variety of industries, especially clothing factories. Farther north the scene is more rural until, near the junction with the Mohawk Valley, are the towns of *Albany*, the capital of New York State and an inland port, *Schenectady*, which has a large electrical engineering industry, and *Troy*, which has clothing and engineering works.

Philadelphia (2 002 000) lies on a peninsula between the Delaware and Schuylkill Rivers, and is a major port and industrial centre. Textile manufacturing is still the leading industry, but it has declined in importance in face of competition from the South. On the other hand, metallurgical industries are increasing not only in Philadelphia but in the Delaware Valley area as a whole. There are large steelworks which make use of iron ore imported from Venezuela and the Labrador–Quebec region, and for this reason they lie close to waterways. Raw steel is in demand locally for use in engineering and shipbuilding industries. There are oil refineries in the Delaware Estuary, using crude oil from Venezuela and the Gulf of Mexico, and petrochemical industries have developed here also. *Wilmington* (250 000) is mainly engaged in the chemical industry.

Fairchild Aerial Surveys Inc.

Bayonne
One of the specialist dock areas in the New York conurbation. Describe the scene.

Baltimore (940 000) is located on the shores of Chesapeake Bay, 240 kilometres from the Atlantic, although ocean-going vessels can reach it via the Chesapeake and Delaware Canal. The growth of the city has benefited from its position not only in the heart of the industrial north-east but also in relation to the southern Atlantic states and the mid-west. Within its industrial orbit is a large steelworks at Sparrow's Point, which imports iron ore from Venezuela and coke from Pennsylvania. Baltimore itself has many major engineering works and ship repairing yards.

Washington D.C. (765 000) lies at the head of the estuary of the Potomac. George Washington selected the site for the capital in 1790, at which time it occupied a central position within the thirteen eastern states making up the United States. Its primary function is as a centre of government and administration, but a number of industries have developed including printing and electrical engineering. D.C. stands for District of Columbia, which is federal territory.

U.S. Information Service

Washington D.C.
The domed Capitol building is where Congress sits. The other large buildings are offices of the Federal Government.

Exercises

Answer in note form:
1. Describe the physical features and soils of the Mid-Atlantic coastlands.
2. In what ways has agriculture been influenced by local markets in this region?
3. Show by means of an annotated sketch map the importance of the Hudson–Mohawk gap.

Essay:
Why is New York the leading port on the east coast of North America?

73

Aerofilms
A suburban housing estate typical of the many which stretch for miles in the industrial belt of the north east.

13: THE SOUTH ATLANTIC COASTAL PLAIN AND PIEDMONT PLATEAU

This broad coastal plain was formerly a submerged continental shelf which has been raised above sea level by earth movements. Consequently it is of low, monotonous relief and the surface is composed of unconsolidated sediments including sands, gravels and clays. The former coastline is marked by a 'Fall Line' which follows the junction of the plain with the old, hard, crystalline rocks of the Piedmont Plateau. Here many streams plunge over the edge of the plateau in a series of falls and rapids.

Much of Florida consists of limestone and there are karstic features including thousands of sink holes which have been formed by the collapse of the roofs of underground caverns. Because of the high water table many of the sink holes contain lakes.

Along the coast the post-glacial submergence has produced numerous estuaries, and there has been a remarkable growth of sand spits, lagoons and marshes, many of the latter being in the process of drying out to become land. One of the largest swamp areas is the Dismal Swamp of east Virginia and North Carolina which contains many fresh water lakes, and an even larger one is the Everglades of southern Florida. In the extreme south of Florida is a broken coral reef known as the Florida Keys, ending at Key West.

The climate of the region is of warm temperate eastern margin type, merging into the sub-tropical in Florida. Summers are hot and humid, and winters are mild, although in winter there is a danger of cold waves surging from the north to bring frosts even into Florida. There is a plentiful rainfall of 1200 to 1500 millimetres per annum, with a summer maximum. Hatteras, which is fully exposed to Atlantic influences including the warming effect of the Gulf Stream, has a January mean temperature of 8°C, a July mean of 26°C, and an annual rainfall of 1397 millimetres.

Sometimes in summer the south is affected by violent hurricanes, which are storms associated with intense low pressure systems round which the winds revolve with great force, causing serious damage and occasional loss of life. In recent years, because of the earlier detection of hurricanes by the use of artificial satellites, there has been more time to evacuate threatened areas, so reducing the loss of life.

Farming

The nearly level land, generally fertile soils and warm, moist climate favour a variety of farming activities. The early colonists earned a living by growing cotton and tobacco, and introduced slave labour, but this proved inefficient and the land gradually became exhausted where it was used year after year for the same crop.

The region remains part of the 'cotton belt' which stretches across into Texas, but cotton is no longer the leading money earner. The area under cotton has declined considerably although yields per hectare have risen as a result of the greater use of fertilisers, the restriction of the crop to the richer lands, and the introduction of improved varieties. Cotton is now rarely grown close to the coast because of the ravages of the boll weevil, a beetle whose grub feeds on the cotton boll, and which is very difficult to eradicate.

Tobacco is a major money crop in the northern parts of the region, particularly in North Carolina, although Virginia and South Carolina

Harvesting tobacco *U.S. Information Service*

75

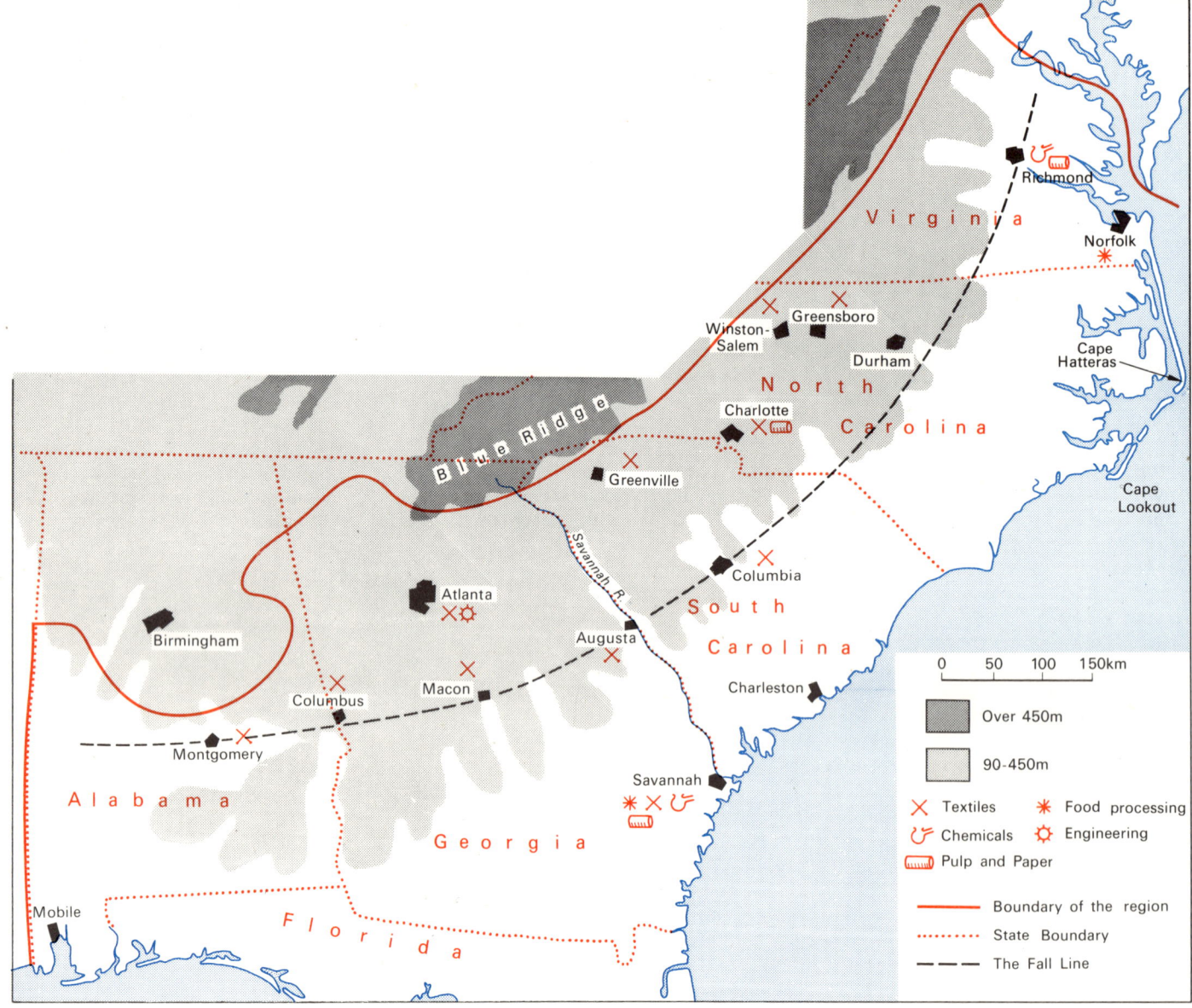

FIG. 13.1 **The South Atlantic Coastal Plain and the Piedmont Plateau**

are also important producers. Like cotton, tobacco soon exhausts the land, and the quality of the crop depends largely on the character of the soil. The main tobacco manufacturing centres are Durham in North Carolina and Richmond in Virginia, where the leaf is stored for from two to four years after curing.

There has been a marked increase in the cultivation of peanuts and in some places near the coast it has replaced cotton as the chief crop. It does best on well drained, sandy loams, and is grown mainly for fodder or for vegetable oil. Cultivation is becoming increasingly mechanised except for harvesting, which requires a lot of labour. Soya beans are another crop which is now of significance especially on the Piedmont Plateau. The crop can either be grazed or the beans can be crushed to provide oil which is used in the manufacture of foodstuffs, soaps and varnishes.

The greatest change in farming in recent years has been the enormous increase in the keeping of livestock. There is no shortage of

U.S. Information Service

This scene is typical of much of the recently reclaimed area of central Florida. A combination drainage-irrigation system is being developed here which draws off water in the rainy season and holds it back in dry periods. The land will be used either for beef cattle or the production of vegetables.

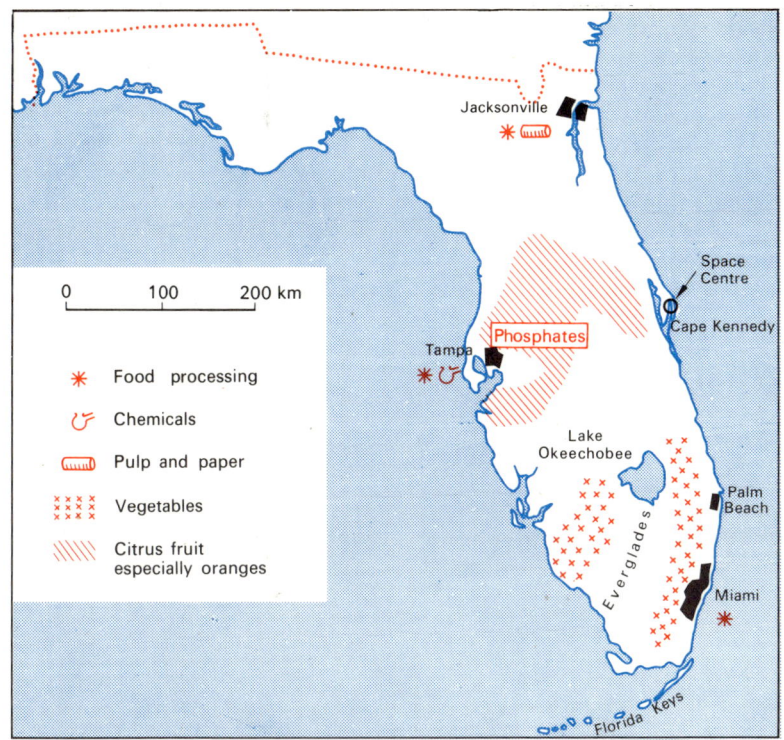

FIG. 13.2 **Florida**

land, and the moist climate and naturally rich soils produce a lush grass which is suitable for cattle. Citrus pulp, sugar cane molasses, peanuts and soya beans provide additional fodder. Central Florida has become noted for cattle fattening, whilst Georgia, Alabama and the Carolinas have registered large increases in pig farming. Half the American broiler chicken industry is located in the region, much of it in agricultural areas which became depressed after the decline of cotton growing.

Florida is particularly well known for market gardening and fruit production. Many of the holdings are large for this form of intensive farming and are highly mechanised. Tomatoes and potatoes are widely grown, and on the deep soils of the Everglades, which are rich in organic matter, beans, celery, sweet corn and cabbage do well.

Rapid road and rail transport as well as a position nearer to the industrial north-east allows Florida to compete successfully with California in the marketing of these crops.

Citrus orchards have expanded greatly in Florida in spite of periodic frosts which have cut back the acreage from time to time. Large scale production is concentrated on the rolling limestone country in central Florida and along narrow coastal strips. Most important are oranges, followed by grapefruit, tangerines and limes. Two-thirds of the crop is crushed to make fruit juice and a lot of this is frozen. There is some growing of sugar cane in the area round Lake Okeechobee, using mechanised methods, but crop failures are liable to occur from time to time because the climatic conditions are very marginal for successful production.

U.S. Information Service
An orange grove in central Florida. What is likely to happen to the fruit shown here?

Forest Industries and Fishing

Over half the region is forested and the southern pine is the most valuable timber, but there has been a decline in demand due to competition from softwoods coming in from western U.S.A. where the trees are larger and the industry is organised on a bigger scale. Formerly the southern pine was much used in house construction, but now its main use is for pulpwood which is made into cardboard and brown paper. Florida is the leading state for pulp but Georgia and Alabama are also important. In addition, these states produce about half the world's turpentine. The swamp areas of Florida provide cypress timber which is used in boats and for other constructional work.

There is some fishing along the shallow Atlantic coast, the catch being of great variety with shrimps and oysters outstanding, but the industry is of declining importance as a result of overfishing and pollution.

Minerals and Manufacturing

About 30 per cent of the world output of phosphates comes from west-central Florida. The phosphates are worked by great dragline diggers and production costs are low. They are used in the manufacture of fertiliser and also in food processing and insecticides.

There are two important manufacturing areas. The first is on the Fall Line where the towns have the advantage of water power and can also receive coal from the west. Textile manufacture is the main industry, especially in the Carolinas where cotton, synthetic fibres and even woollen industries are well represented. Local resources of timber and tobacco have also given rise to manufacturing here.

The second area is Florida, where industrial expansion has been even more remarkable. The main reason for this development is the attractive environment which brings people in to provide the necessary labour. Woollen and clothing factories are widely spread through much of the state. The manufacture of rayon has benefited from the availability of local cotton linters and wood pulp, whilst nearby chemical plants provide the necessary acids. There are many large, new engineering works, mostly concerned with electrical engineering and aircraft production. The rocket launching base at Cape Kennedy has stimulated these developments.

The Holiday Industry

The holiday industry is well developed along the coast especially in Florida, to which thousands come to escape from the bitter winters

78

A phosphate plant near Barstow, Florida
About three-quarters of the nation's production is obtained from this area.

of the north-east and to enjoy the bathing, boating and fishing. The cypress swamps of the Everglades are an additional attraction, and there are many large resorts including Miami and Palm Beach. Holiday centres have also grown up along the shores of Georgia and the Carolinas; and Virginia receives numerous visitors, partly on account of its many historic associations.

Towns

Some of the first towns to develop were on the Fall Line, which marks the limit of navigation of the rivers. The falls and rapids were utilised to drive the early machinery and now they are harnessed to generate hydro-electric power. Fall Line towns include *Montgomery*, Alabama, noted for its cotton and livestock market, and *Richmond*, Virginia, which has textile, pulp and paper, and tobacco industries. *Atlanta*, Georgia (550 000) is a rapidly expanding commercial and industrial centre on the Piedmont Plateau producing textiles, furniture and machinery.

Most other large towns lie on the coast and these include *Mobile*, Alabama, a port with an oil refinery and chemical, pulp and paper, and textile industries, *Savannah*, Georgia, and *Norfolk*, Virginia (305 000), both the latter being ports and industrial towns. In Florida the chief towns are *Miami* (380 000), the holiday capital, and *Jacksonville* (200 000), a commercial centre noted for timber interests.

Exercises

Answer in note form :
1. Describe and explain the chief characteristics of the climate of the South Atlantic states.
2. What changes have taken place in agriculture in recent years in this region?
3. Write an account of the holiday industry in Florida.
Essay :
Locate the main manufacturing industries of the South Atlantic states and explain how geographical factors have encouraged their growth.

A highly automated coal pier at Norfolk, Virginia which handles much of the Appalachian coalfield's exports to over 50 foreign countries.

14: THE APPALACHIANS

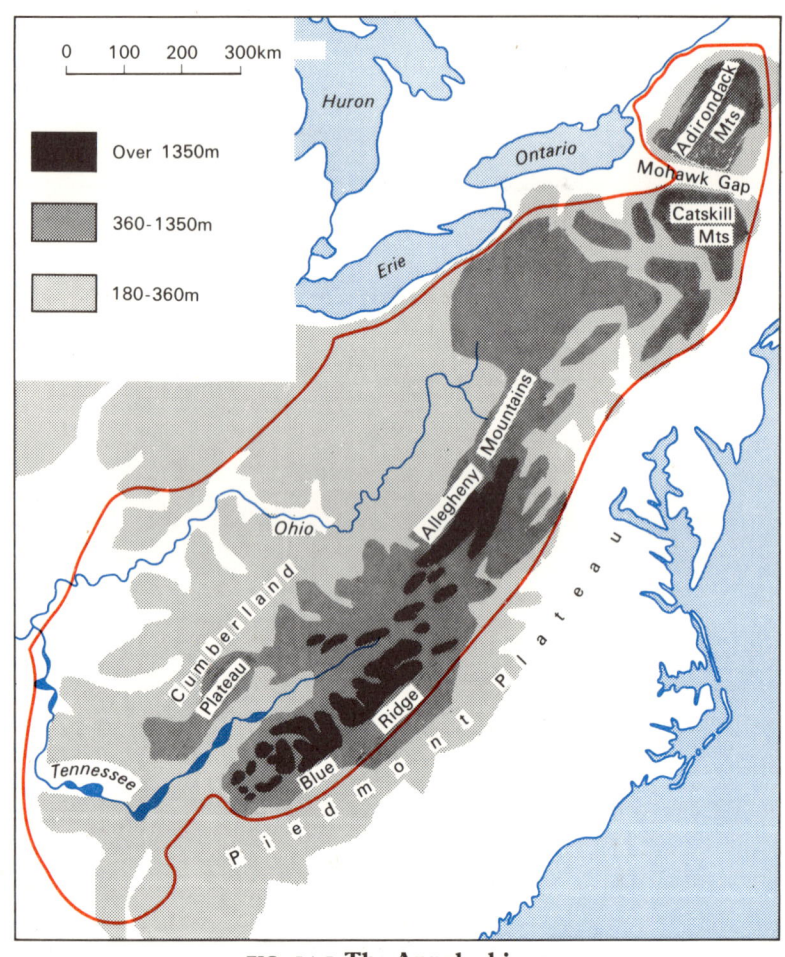

FIG. 14.1 **The Appalachians**

Fairchild Aerial Surveys Inc.

The wooded ridges of the Blue Mountains.

The Appalachians are a complex system of ridges, deep valleys and plateaus. Its rocks were subjected to early folding, followed by a period during which the mountains were worn down to a peneplain, then uplifted again and further dissected by rivers.

The main features of the region are shown in Figures 14.1 and 14.2. The Blue Ridge, which consists of ancient igneous and metamorphic rocks, is the dominant feature. In the north it forms a series of smooth ridges rising to 900 metres above sea level, whilst in the south the ridges are more broken but reach 1800 metres in places.

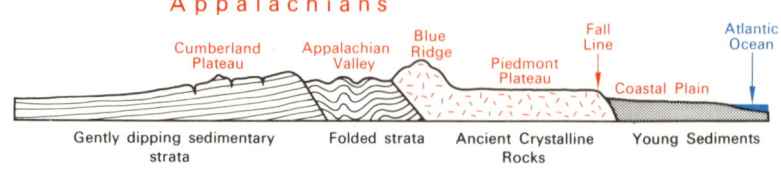

FIG. 14.2 **Section from north-west to south-east across the Southern Appalachians**

U.S. Information Service

An old photo showing gulleying caused by heavy rains on recently cleared land in the Tennessee Valley.

To the west lies a belt of folded sedimentary rocks, with ridges of limestone and sandstone following the south-west to north-east Appalachian trend. A prominent feature here is the Appalachian Valley, which is really a wide trough following a zone of less resistant rock and containing a number of parallel valleys. Farther west still lies an extensive dissected plateau formed of almost horizontally bedded sedimentary rocks, known as the Allegheny Plateau in the north and the Cumberland Plateau in the south.

As might be expected, the climate shows considerable variations depending on both altitude and latitude. In general, winter temperatures are much lower than along the Atlantic coastal plain and there are short, bitterly cold spells. Summers are warm and rainfall is plentiful, ranging from 1000 millimetres per annum in lower areas to over 1500 millimetres on some of the higher land.

Many of the slopes remain under forest which consists both of conifers, including spruce and fir, and deciduous trees such as oak,

beech and maple, but the more accessible timber has been cleared, often in the past by very wasteful methods. Forest products include some constructional timber, pulp and paper.

Farming

The soils are generally of poor quality so that farming in the past has been little more than of a subsistence type. The low standard of living was also partly due to the fact that many of the early settlers came here from the richer lands seeking a retreat from competition from more ambitious and hard-working neighbours. Thus the label of 'hillbillies' for these people was true for much of the nineteenth and early twentieth centuries.

Deforestation and the growing of crops like cotton and maize which do not completely cover the soil have led to serious soil erosion in the Appalachian region. The impoverished soil, without its natural protection, has been slowly carried away by the heavy showers of rain, and deep gullies have been cut on the hill slopes. The situation was most serious in the Tennessee Valley where almost the whole population had become impoverished. To improve the economy of the area, the United States Government set up the Tennessee Valley Authority in 1933. It acted in a variety of ways, one of the most important being the construction of 17 dams on the Tennessee and its tributaries, in order to control flooding and improve navigation. The dams are a source of hydro-electricity, some of which is used in new industrial

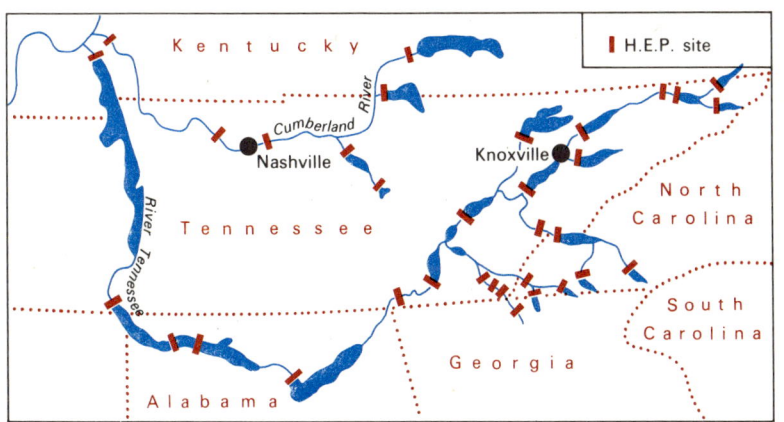

FIG. 14.3 **Hydro-electric power sites in the Tennessee Valley area**

Ewing Galloway, N.Y.

In what way has the construction of dams such as this, the Chicka-maugua on the Tennessee, aided development in this region?

U.S. Information Service

Strip rotation and contour ploughing are a check on both wind and sheet water erosion.

U.S. Information Service

To combat further erosion check dams and newly planted red pine trees will help.

undertakings including the production of fertilisers and agricultural machinery. It has also been necessary to plug the gullies and to anchor the soil by establishing tree belts. Farmers have been encouraged to practise contour ploughing and strip cultivation, with a greater variety of crops.

In the northern Appalachians, in the states of New York, Pennsylvania and Ohio, dairying is carried on to supply produce to the nearby towns. The farms are small, normally having about 20 to 25 cows, and the land is partly pasture and partly under fodder crops such as oats and maize. Some farmers specialise in poultry rearing and many keep sheep. These provide good quality wool and lambs are raised for meat.

The south and central part of the Appalachians have more mixed purpose farms concerned with livestock, cereals and fruit. The richest part is the bluegrass area of Kentucky, which is noted for tobacco growing. This tobacco area makes Kentucky second to North Carolina in production although the crop occupies only a very small acreage. Yields are heavy as the soil is alluvial and has developed on limestone with a high phosphate content. It is this latter which gives the grass its blueish colour.

Mineral Working and Manufacturing

On the western flanks of the Appalachians is the most productive coalfield in the world, and this was a major influence on the industrial growth of the United States in the nineteenth century. The coal is of bituminous type and the main producing areas are in West Virginia, Pennsylvania and Kentucky. Seams are near the surface, very thick and relatively undisturbed by folding or faulting. Most of the coal is now cut from the face, loaded and carried by conveyors automatically. There has also been an increase in open cast mining, the method being to remove the overburden and dig out the coal by power shovels. Coal from the Appalachian field is carried mainly by rail and the biggest users are electricity power stations, coke converters and domestic heating.

Nearly all America's anthracite is produced from mines in north-east Pennsylvania. Mining here is difficult since seams are thin, broken and steeply inclined, but even so most mining processes are mechanised. The main use of the anthracite is for household heating and a big advantage is its nearness to the great cities of the north-east. However, there has been a marked drop in production in recent years due to competition from oil and gas central heating.

It was at the end of the nineteenth century that the great American oil industry started in north-west Pennsylvania. For some time this was the most important field, but production is now small, although the oil is very pure and is ideal for lubricants. Natural gas is far more valuable and the gas field extends from Pennsylvania into West Virginia and Kentucky.

On the edges of the Appalachians are a number of major industrial areas. Some, such as the Pittsburgh, Youngstown and Scranton areas, developed in the nineteenth century and owe their importance mainly to the coalfield. Other areas, notably along the Ohio, have expanded, particularly in the twentieth century.

On the anthracite field is a densely populated area including Scranton and Wilkes-Barre. Here, although mining has declined, there is much industrial activity of which engineering is the most important.

Pittsburgh (610 000) developed as a metallurgical centre at the junction of the Allegheny and Monongahela Rivers where coal, iron ore and limestone could be cheaply assembled. The local Connellsville coal produces fine quality coke, and now that the local iron ores are worked out, Lake Superior ores can be brought in by way of the Great Lakes waterways. Today the Pittsburgh district is the second most important steel producing area in the United States, after Chicago. A great variety of rolled steel products are made in mills which sprawl for almost 120 kilometres along the Allegheny, Monongahela and Ohio Rivers. Pittsburgh has numerous engineering works including the vast Westinghouse electric firm, and it is also the home of Heinz foods.

To the north-west of Pittsburgh is the *Youngstown* district, the third

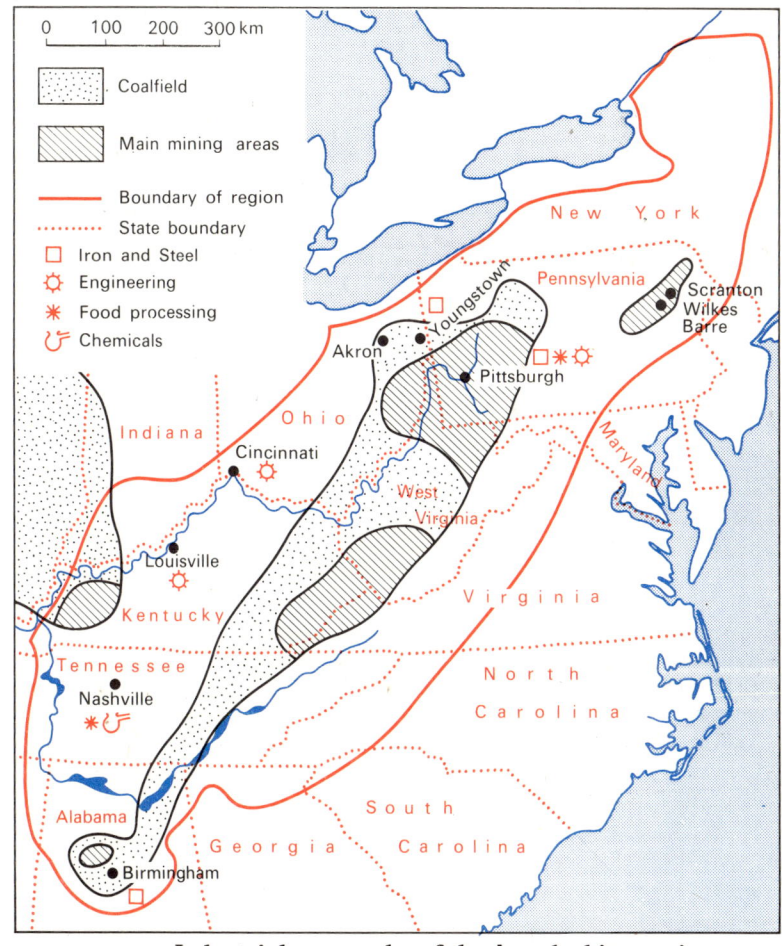

FIG. 14.4 **Industrial geography of the Appalachian region**

Pittsburgh *U.S. Information Service*

most important steel producing area, specialising in thin sheets of steel and other components for the car industry. Nearby is *Akron* which also has links with the car industry since it is dominated by rubber industries, especially the manufacture of tyres.

Along the banks of the Ohio River are a series of industrial towns which owe their importance to local coal and salt reserves, the latter having led to the development of chemical industries. There are numerous electricity power stations in the area using coal for fuel, and both steel and aluminium are produced in large quantities.

Cincinnati (500 000) is the main focus of river and rail routes on the Ohio. It grew up as a river-side settlement but has now spread on to the surrounding hills, its industries being chiefly concerned with engineering, including the production of machine tools, cars and aircraft. Farther downstream is *Louisville* (390 000), which developed near the Ohio Falls at a point where goods carried by river had to be offloaded, and is now a great market and distributing centre. Among its industries is a large General Electric Company factory which produces household electrical goods. Farther south is *Nashville,* the commercial and main route centre of Tennessee, which has food processing and chemical industries.

In the extreme south of the region is another industrial area centred on *Birmingham* (340 000), Alabama. Here local iron ore, coal and limestone have led to the development of a considerable steel industry whose many mills produce a wide range of finished steel goods. Today the local ore is supplemented by imports from Venezuela.

Exercises

Answer in note form:

1. Describe the physical features that would be met with in a journey across the Appalachians from east to west.
2. What were the reasons for the setting up of the Tennessee Valley Authority and what did it achieve?
3. Make notes on the Appalachian coalfield, mentioning its location, coal production and uses made of the coal (reference may be made to Chapter 10).

Essay:

Write an ordered geographical account of the industries of the Appalachian region.

15: THE GREAT LAKES REGION

The boundary between the United States and Canada passes through Lakes Superior, Huron, Erie and Ontario. This leaves the western and southern shores of Lake Superior, the whole of Lake Michigan, the western shore of Lake Huron, and most of the southern shores of Lakes Erie and Ontario within the United States.

The origin of the Great Lakes and their use as an inland waterway have already been dealt with on pages 7 and 35. As in Canada, much of the land near the lakes is covered by glacial drift, consisting of

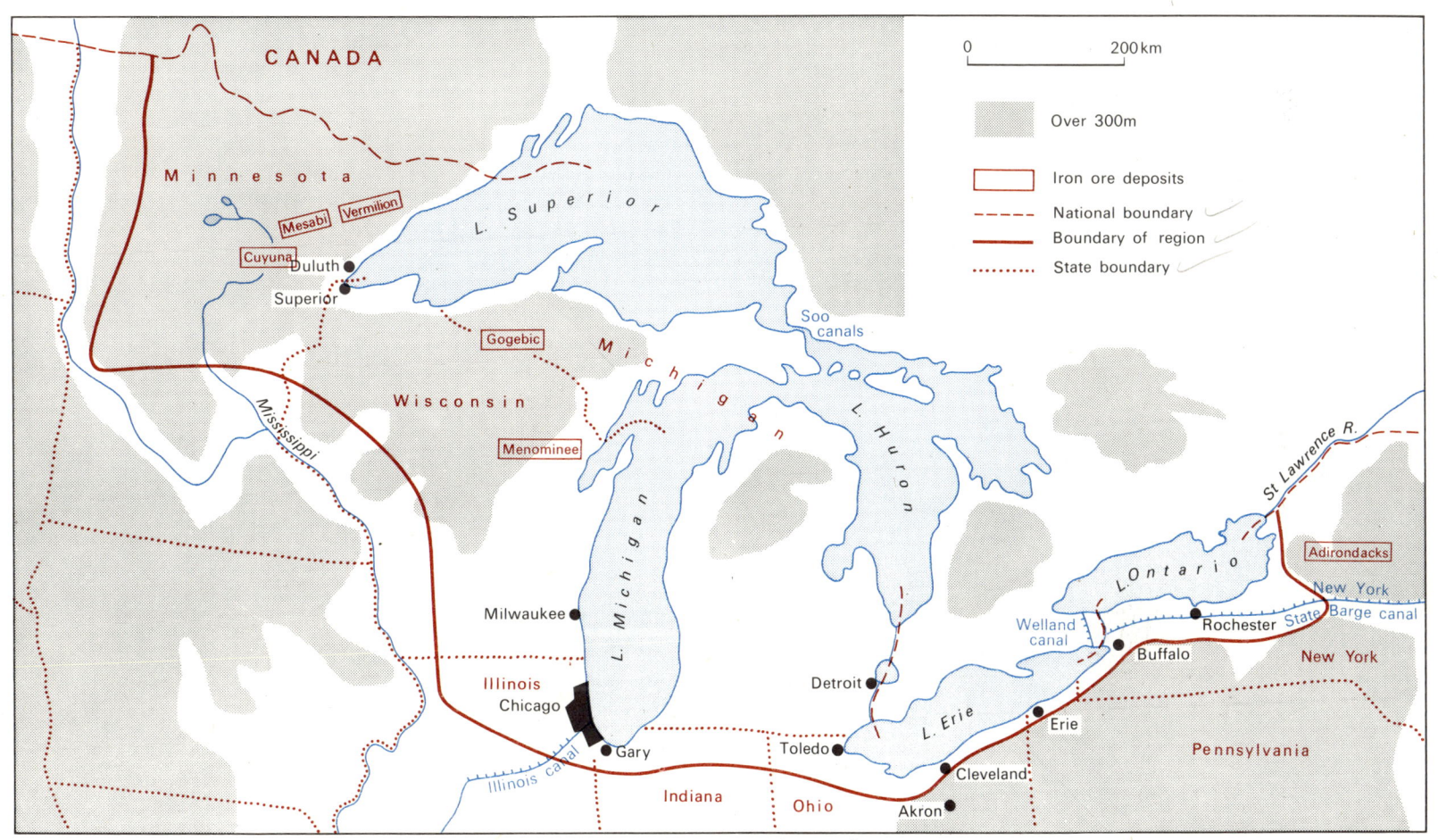

FIG. 15.1 **The Great Lakes region**

sands, gravels and clays, but the drift is more continuous and thicker to the east of Lake Michigan than to the west.

In the main, the climate is governed by continental influences. Winter conditions are severe with average temperatures about or below freezing point for at least three months. Summers are fairly hot and most rainfall occurs then. The amount of rain decreases from some 1000 millimetres per annum in the east to 650 millimetres in the west. Chicago, whose climate is fairly typical of the region, has a mean January temperature of $-3°C$, a July mean of $23°C$, and an average rainfall of 820 millimetres.

The natural vegetation is mixed deciduous and coniferous woodland, but there was indiscriminate tree felling during the nineteenth century, and most of the existing woodlands have grown up in the present century. Lumbering is carried on in the western parts of the region and a considerable pulp and paper industry has developed here.

Farming

The type of farming has been influenced by the relatively short growing season, the plentiful summer rainfall, and the large urban markets nearby. Mixed farming, with an emphasis on dairying, is the most profitable type, and for this reason the region is known as the 'Dairy Belt'. Dairy farms are small and much of the land is used to produce fodder crops such as hay, oats, alfalfa and maize. The latter does not have a long enough growing season to mature fully so that it is cut green and is very nutritious fodder for dairy cattle. Wisconsin has heavy clay soils which produce rich grassland and is the leading dairy state; much of its milk goes to the Chicago and Detroit areas, and there is a large production of cheese. Other important dairy states are Minnesota, which is noted for butter, and Michigan.

On the deep, fertile glacial drift which forms a narrow belt south of Lake Erie, dairying is still the dominant activity but farming is more varied. Maize, wheat, oats and soya beans are important crops, and there is fruit growing, including apples, cherries and even peaches and grapes. However, the principal area for fruit is to the east of Lake Michigan. The influence of the nearby lakes gives a more equable climate, which is one reason for this specialisation on fruit growing.

Mineral Working and Manufacturing

The significance of the Appalachian coalfield to the American economy has already been stressed, but equally important are the massive

U.S. Information Service

Minnesota is still primarily a rural state with dairy farms, woodlands and small villages.

Fairchild Aerial Surveys Inc.

Iron ore workings at Hibbing in the Mesabi Range.

A lake freighter proceeding through St Mary's Locks at Sault Ste Marie, Michigan. What might it be carrying?

iron ore deposits round Lake Superior. The deposits occur widely round the shores of the lake but the Mesabi Range, Minnesota, which was first developed in 1892, has dominated production. The best ore is a high grade haematite, with an average metal content of 64 per cent, and mining operations are relatively simple. Electric shovels dig out the reddish ore from vast open pits and load it on to conveyor belts. These transfer it to preparation plants which crush and wash the ore ready for shipment from various Lake Superior ports. So much ore has already been dug out that reserves available to a number of iron and steel companies are becoming rapidly worked out. However, there are large amounts of low grade ore with a metal content of about 30 per cent in the Mesabi and elsewhere, and new equipment has been developed to concentrate the ore before shipment. Iron ore from the Adirondacks is also of considerable importance.

Most of the iron ore from the Superior fields is shipped from Duluth and Superior to ports on Lake Erie and to Chicago. *Duluth* has an iron and steel industry of its own, using local ores and Appalachian coal. Its ore shipments and the large quantities of grain and timber which it sends down the Great Lakes make it one of the biggest ports in the United States.

Whilst the western portion of the American Great Lakes region is concerned especially with primary production – farming, lumbering and mining – the eastern or lower Great Lakes region is mainly a manufacturing area. The reasons for this are: (a) the proximity of the north Appalachian and Illinois coalfields; (b) the Great Lakes–St Lawrence waterway system which provides cheap transport for Lake Superior iron ore and the increasing amounts of imported ore; (c) the large local populations which provide markets and labour; and (d) lakeside sites with plenty of room for expansion.

Metallurgy is the basic industry and employs over half the working population. The expansion of the steel industry has resulted from the growth of steel consuming industries such as the manufacture of motor cars, railway equipment and machinery.

Chicago (3 600 000), the largest city of the region, lies at the southern end of Lake Michigan. It grew up at the mouth of a small creek where a glacial overflow channel leads to the Illinois River, and this is now followed by a canal, roads and railways. It occupies a nodal position and is the great commercial centre of the mid-west, a market for much of the northern Mississippi basin and a distribution centre. This is why the biggest mail order firms in the country have their headquarters here.

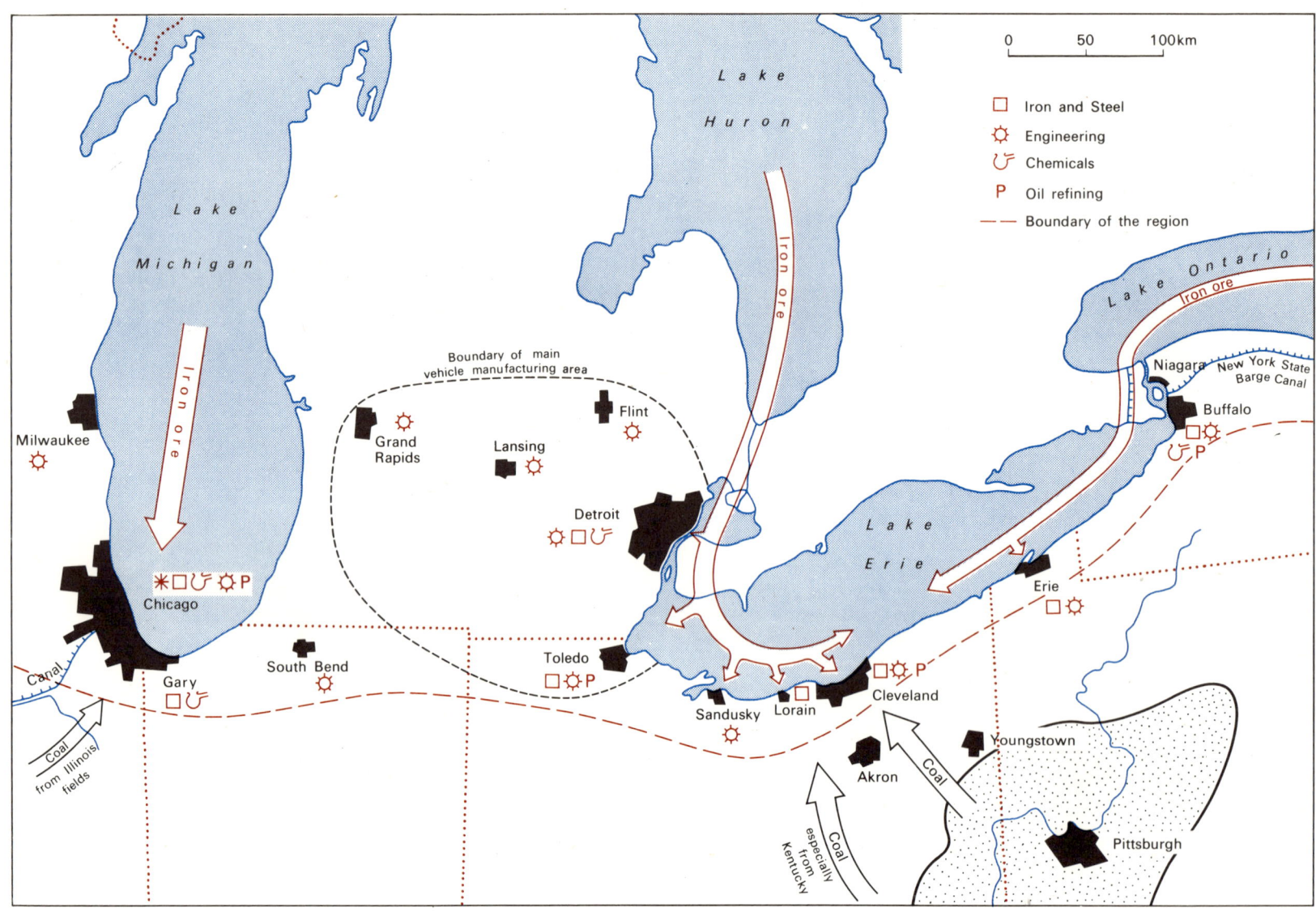

FIG. 15.2 **The industrial belt of the Great Lakes region**

Chicago   *U.S. Information Service*

Ewing Galloway N.Y.
Steel mills, docks and car works, all part of the Ford plant at Detroit.

Above all, Chicago is a manufacturing city. Its position on the edge of the maize belt has made it a meat packing centre dealing with both pigs and cattle. The expansion of oil refining has encouraged the growth of chemical and associated industries such as the manufacture of soaps and detergents. There are many branches of engineering including the production of railway rolling stock and agricultural machinery.

The Chicago area is the leading steel producing region in the United States. The U.S. Steel Corporation's works at Gary are capable of producing over seven million tonnes of steel per year. They were built on waste land on the southern shore of the lake and have attracted engineering industries to the area.

Milwaukee (750 000), on the western shore of Lake Michigan, concentrates on the manufacture of generators and other equipment used in electricity power stations.

Detroit (1 750 000) lies on a narrow neck of land between Lakes Huron and Erie, where the land route from the Lakes Peninsula of Ontario crosses the water route. It owes its recent growth to the motor car industry, established here due to the early efforts of Henry Ford who was born nearby. There were traditional wooden carriage works in the area which had some influence, but it was the availability of steel that made possible the rapid expansion of the industry. Today production is dominated by three large car firms – General Motors, Ford and Chrysler – but there are also a few independent companies. Assembly line methods are employed, with large numbers of separate factories producing the necessary components.

Many of the towns on the southern shores of Lake Erie originated as lake ports handling iron ore from Lake Superior and then established their own metallurgical industries. *Toledo* is such a port, now concerned especially with the car industry and oil refining. The outstanding Lake Erie city is *Cleveland* (876 000) which is sited at the mouth of a river giving it easy access to the coal of the Ohio Valley. Most of the iron ore it now receives is from Labrador and it has varied metallurgical industries including the production of steel, cars and aircraft, and precision engineering. The *Buffalo–Niagara conurbation* (1 054 000), at the junction of the New York State Barge Canal and Lake Erie, is a focus of routes funnelling in from the west into the Mohawk valley. Besides the inevitable metallurgical industries, chemical manufacture is important partly as a result of the availability of hydro-electric power from Niagara Falls.

Farther east are several centres with more varied activities. *Rochester*, for example, has specialised factories producing photographic, scientific and optical equipment.

Exercises

Answer in note form :
1. What are the main types of farming in the Great Lakes region and why are they found there?
2. Draw an annotated sketch map to illustrate the position and importance of Chicago.
3. Describe the location and character of the motor car industry in the Great Lakes region.

Essay :
Explain how geographical factors have encouraged the development of manufacturing industries along the shores of the Great Lakes.

16: THE CENTRAL MISSISSIPPI LOWLANDS

The Mississippi is one of the world's greatest rivers, and among its many tributaries are the Red River, Arkansas, Ohio and Missouri, the latter being considerably longer than the Mississippi itself. This chapter deals only with the central part of the Mississippi basin, i.e. the area between the Great Lakes region and the Gulf Coast lowlands.

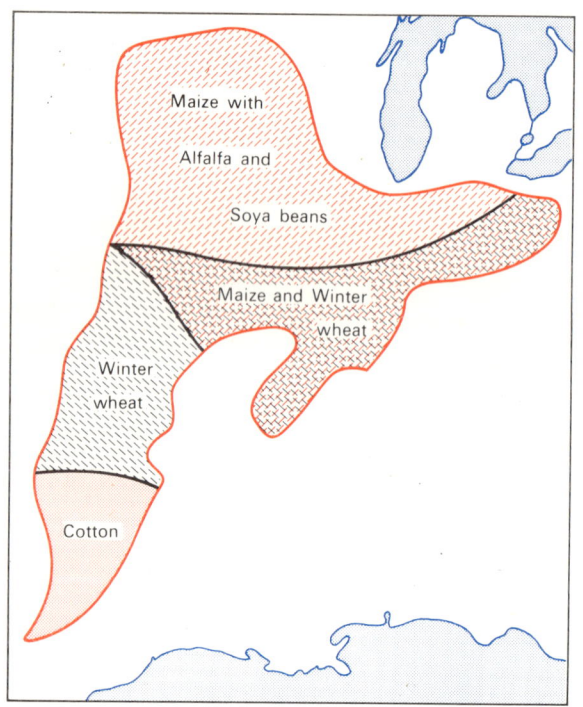

FIG. 16.1 **The Central Mississippi Lowlands – agriculture**

It is a vast, undulating lowland, formed of relatively undisturbed sedimentary rocks including sandstones, limestones and conglomerates. North of the Missouri and Ohio much of the land is covered by glacial drift, with widespread areas of boulder clay as well as gravel and sand, but to the south of these rivers the land is unglaciated. Thus there is considerable variation of soil, but in general it is of the rich, dark prairie type, formed in a similar way to the prairie soils of Canada (page 46).

In spite of the southerly latitude, the climate is subject to continental influences. Summers are hot and winters fairly severe. The annual rainfall varies from about 1000 millimetres in the east to only 500 millimetres along the western boundary, and most of it falls in summer, in the form of heavy showers. St Louis, near the centre of the region, has a mean January temperature of 0°C, a mean July temperature of 26°C, and an annual rainfall of 1000 millimetres.

Farming

Maize is the chief crop of the region. It thrives where there is a rich organic soil and a frost-free period of at least 150 days. The moist spring for rapid germination, summer showers, and hot dry September for ripening and harvesting, have all helped to make this the world's most productive maize region.

U.S. Information Service

Harvesting Maize.

U.S. Information Service

Hogs are an important part of the economy in the Maize belt. This is a scene in Missouri.

Most of the maize is consumed on the farm by livestock. Half of it is fed to pigs and two-thirds of America's pigs are kept here, particularly in east Iowa and north-west Illinois. Cattle are brought from the western high plains, mainly in October, for fattening, which takes from seven to eight months. Most farms have some poultry, kept mainly for eggs, but the broiler business is now expanding.

Maize is grown in a rotation because it is an exhaustive crop. Oats and clover planted together fit well into the rotation. After the oats are harvested the clover continues to grow and is frequently ploughed in to enrich the soil. Alfalfa is also grown as a fodder crop and sometimes replaces clover in the rotation.

Many farmers are turning to soya beans as an alternative to maize. The crop can be grazed or harvested for use as winter feed for animals, but it has not yet displaced maize as the principal crop of the region. Winter wheat is another major crop especially in the southern areas.

Mineral Working

The deposits of both coal and oil in this region, although overshadowed by those in neighbouring areas, are nevertheless of considerable significance. There are a number of scattered coal mining areas associated with the extensive eastern and western interior coalfields. The Illinois fields are the most productive and include valuable seams of coking coal in the south. Illinois is also fortunate in having quite considerable oilfields although production is only about one-fifteenth of that of Texas. In addition, a small part of the great interior oilfield of Kansas/Oklahoma is located in this region.

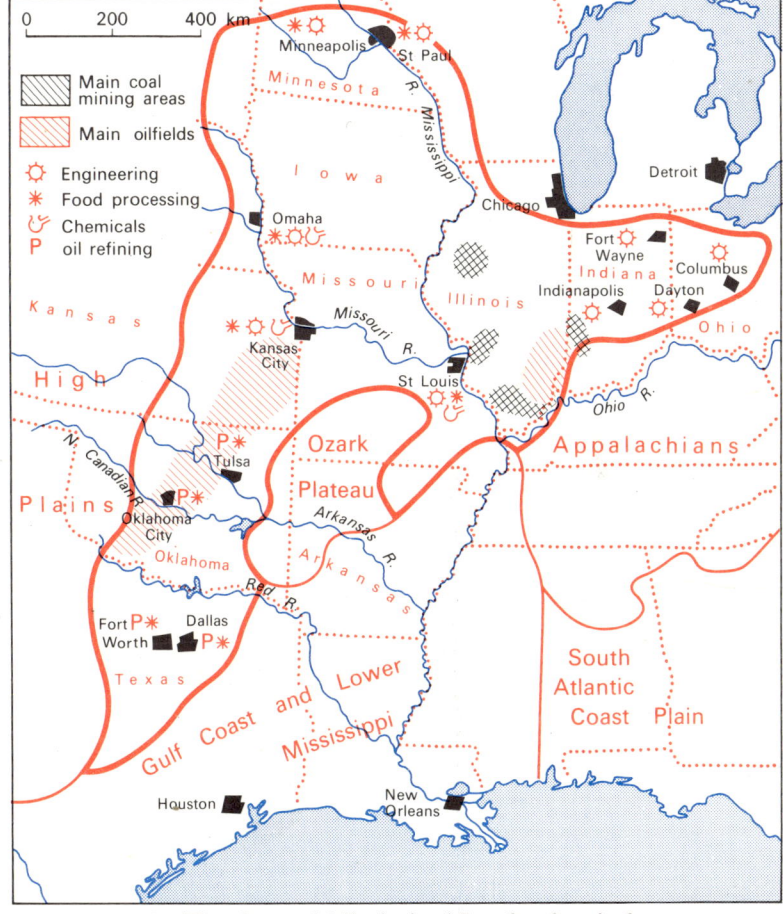

FIG. 16.2 **The Central Mississippi Lowlands – industry**

Towns

The dozen or so major towns of the Central Lowlands have many features in common. Most are on rivers, often at important crossing points which made them trading stations during the pioneer days. All are now regional centres providing shopping and marketing facilities for the surrounding farmlands, and have a cluster of skyscrapers in the centre marking the commercial district. Many have food processing especially meat packing as a major activity, and those near to oilfields or coalfields have commercial interests in connection with them. Besides this, all have developed other major industries during recent years.

In the north-east are four large towns concentrating on engineering, which is to be expected since they are nearest to the metallurgical centres of the Great Lakes and north Appalachians. Thus *Fort Wayne* is concerned with transformers and wire for the transmission of electricity, *Columbus* (471 000) and *Indianapolis* (476 000) with aircraft manufacture and *Dayton* produces office equipment.

St Louis (2 000 000) is the hub of communications in the centre of the region. It was a fur trading station in pioneering days and became a river port and important gateway to the west. Now it has a broad industrial basis, the main activities being food processing, the manufacture of boots and shoes (based on locally produced leather), chemicals and engineering equipment.

In the west are four pairs of towns. *Dallas* (1 000 000) and *Fort Worth* (500 000) have grown rapidly since the coming of the railways, and both are engaged in the production of cars and aircraft. However, whilst Dallas has links with the oil industry, Fort Worth is a market for farm produce and has meat packing industries. *Oklahoma City* (324 000) and *Tulsa* are also meat packing centres and have strong connections with the oil industry. Tulsa in particular claims to be the oil capital of the world, for it is the headquarters of over 300 oil companies.

Kansas City (476 000) and *Omaha* (301 000) are at either end of an expanding industrial belt. Both towns have agricultural processing industries and Omaha is the country's leading meat packing centre. The demand for fertilisers and insecticides has led to the growth of a chemical industry, and agricultural engineering has developed to supply the needs of this rich farming area. The new industries are more varied in character, with several branches of engineering well represented.

The twin towns of *Minneapolis* and *St Paul* (combined population 1 000 000) grew up at the head of navigation of the Mississippi, just below the Falls of St Anthony which provide hydro-electric power. They remain large communication and market centres both for river and rail borne traffic. Food processing industries are dominant and include flour milling, meat packing and the production of linseed oil, but again new industries have expanded, notably the manufacture of machinery and printing.

U.S. Information Service

Kansas
How does this scene reflect the economy of the state of Kansas?

The Ozark Uplands

This highland area forms a distinct sub-region within the Mississippi basin. Here ancient rocks come to the surface and form a dissected plateau rising to over 600 metres above sea level. The Ozark block is separated by the Arkansas Valley from the Ouachita Mountains to the south, which are of similar structure. Both uplands for long remained isolated, forested and sparsely populated, but there are two areas which have now been considerably developed. One is the Springfield Plateau where farming has been improved and both cattle rearing and dairying are important. The other is the Arkansas Valley which has more varied agricultural activities including the keeping of beef and dairy cattle, the growing of fruit and market gardening.

The thickly wooded upland areas provide a source of timber, such as oak, hickory and pine, for sawmills and pulp mills. More significant are the mineral resources. Lead is mined from underground workings in south-east Missouri and nearby, from north-east Oklahoma and neighbouring states, comes a considerable production of zinc. Even iron ore mining in the Ozarks is on the increase, which is partly due to the working out of the best Lake Superior ores.

Although there is no spectacular scenery in the region, it offers a contrast to the surrounding plains and tourism is increasing with the woodlands and lakes providing the main attractions. Some of the largest lakes are the result of dams built to provide hydro-electric power.

Exercises

Answer in note form:

1. Compare the climate of St Louis with that of Boston and give reasons for the differences.
2. In what ways is the Ozark Plateau distinguishable from the surrounding lowlands?
3. Choose THREE contrasting examples of mid-western cities and explain how and why they are different.

Essay:

Write a full account of farming in the Maize (Corn) Belt, locating the region with a sketch map.

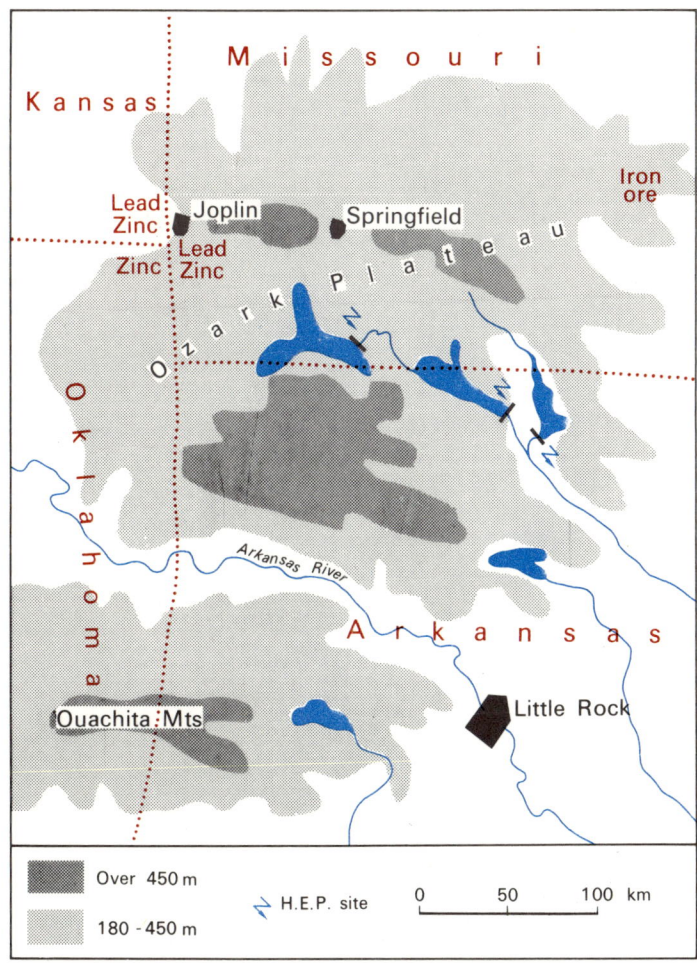

FIG. 16.3 **The Ozark Uplands**

17: THE GULF COAST AND LOWER MISSISSIPPI

This region is a vast lowland stretching from Memphis in the north to the Mexican border in the south-west and Florida in the south-east. The Mississippi and other rivers flowing into the Gulf of Mexico have brought down large quantities of silt and have built up extensive alluvial plains. Along the coast is a belt of marsh over 100 kilometres wide in places. The Mississippi has formed numerous meanders and in the past has changed its course many times, cutting off its meanders to form ox-bow lakes. In some places it flows above the level of the surrounding land, so that levees have had to be built to prevent flooding. The channel has also been straightened at some points, but despite these measures flooding does sometimes occur, especially in spring and early summer. Where it enters the sea the Mississippi has built a large lobate (bird's foot) delta which is constantly growing in size. Two navigable distributaries have been kept narrow so that the current is powerful enough to prevent silting.

The climate is sub-tropical and proximity to the Gulf of Mexico has given the region generally equable conditions. Summers are hot and humid and winters very mild, although there can be short, cold spells and mists are frequent near the rivers because the water flowing from the north is colder than the air. The rainfall amounts to about 1250 millimetres per annum along the lower Mississippi, but decreases to as little as 500 millimetres in the west. New Orleans has a mean January temperature of 12°C, a July mean of 28°C, and an annual rainfall of 1420 millimetres.

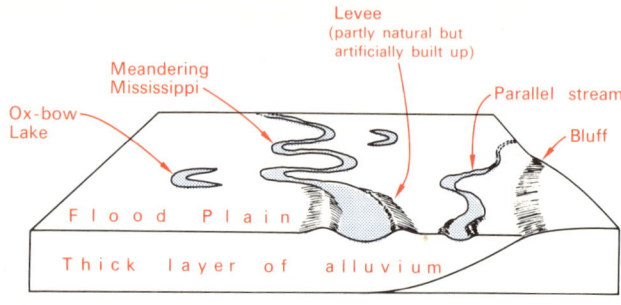

FIG. 17.1 **Block diagram of a portion of the Mississippi Flood Plain**

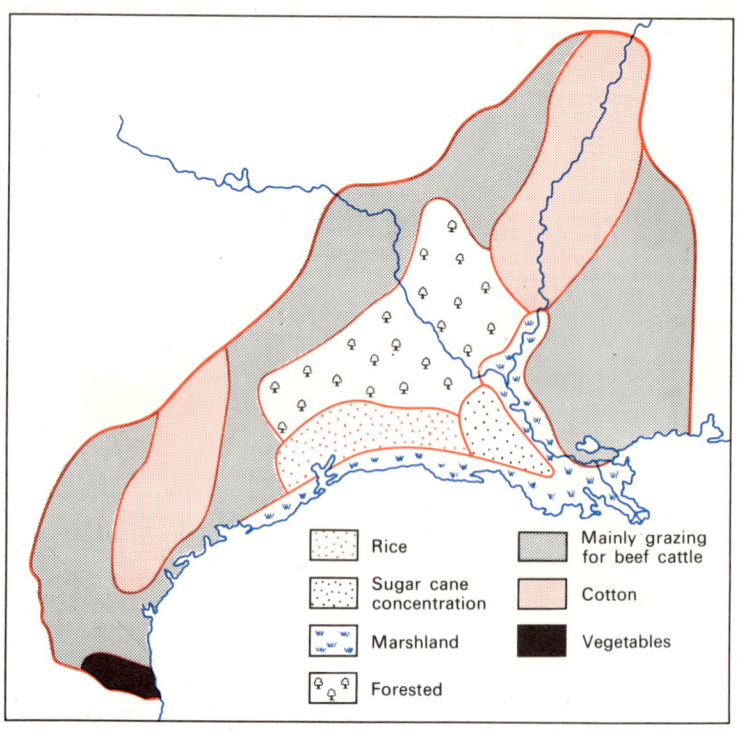

FIG. 17.2 **The Gulf Coast and Lower Mississippi – land use**

U.S. Information Service

Explain the significance of these two photographs.

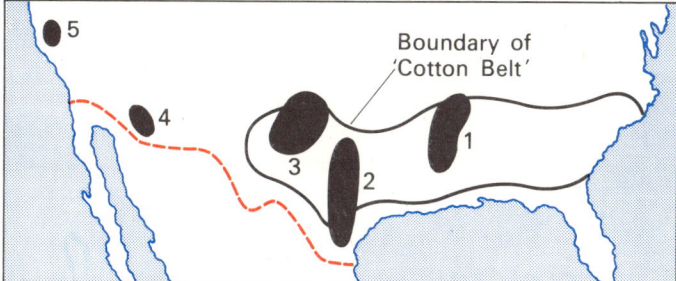

FIG. 17.3 **Cotton growing areas of the United States**
1 Mississippi Flood Plain   4 Irrigated areas of Arizona
2 Black Waxy Prairie of Texas   5 California
3 High Plains of Western Texas

The location of the 'cotton belt' is shown in Figure 17.3. The most suitable conditions for cotton growing are a rich, deep, moisture-retaining soil, a 200 day frost-free period with average temperatures over 25°C for June, July and August, an annual rainfall of between 750 and 1250 millimetres, and dry, sunny weather during the maturing period. Heavy rain immediately prior to picking can spoil the bolls which by then will have burst to disclose the cotton lint.

Two areas have particularly suitable conditions for the crop. These are the Black Waxy Prairie of Texas and the Mississippi flood plain. During the twentieth century the main producing areas have shifted westwards and now include parts of the High Plains of western Texas. Here production is on a large scale and cotton picking machines are employed. In the older cotton lands there are many small holdings, often worked by negroes who are descended from the early slaves, and here picking is generally by hand.

After picking, the bolls are taken to the ginnery where the fibre is separated from the seed. The latter is crushed for oil and the residue is made into cattle cake. American cotton is of medium staple, which means that the average length of the fibre measures about 30 millimetres.

As on the Atlantic coast, the boll weevil has been a great pest, and even today regular spraying is needed in order to control it. Agriculture has become more diversified in recent years, and the acreages under maize, soya beans, vegetables and other crops are increasing steadily.

On the coastal belt, rice is the dominant crop in Louisiana and eastern Texas. Methods of cultivation are highly mechanised, and the rice is now often sown by aeroplane and combine harvested. Water is pumped from the rivers to flood the rice fields. Some sugar cane is also grown in southern Louisiana.

Farther west along the coastal plain in Texas, a greater proportion of land is under pasture for dairy and beef cattle. The 'coastal bend' centred on Corpus Christi is a large dairying area and here cotton and sorghums are also grown. There are pockets of market gardening, especially near the bigger cities, where the crops are often grown under irrigation and benefit from the long growing season. Another rich but isolated farming area is the lower Rio Grande Valley where cotton, fruit and vegetables are grown and dairy cattle and poultry kept.

U.S. Information Service

Harvesting sugar cane in Louisiana.

U.S. Information Service
A Mississippi River towboat pushing 40 barges upstream. What are the advantages and disadvantages of this form of transport?

Along the coast half the land is wooded with pine and various hardwoods, and there are important woodpulp industries. Two minor industries are shrimping, which is carried on in the Port Isabel area, and trapping especially of muskrat in south Louisiana.

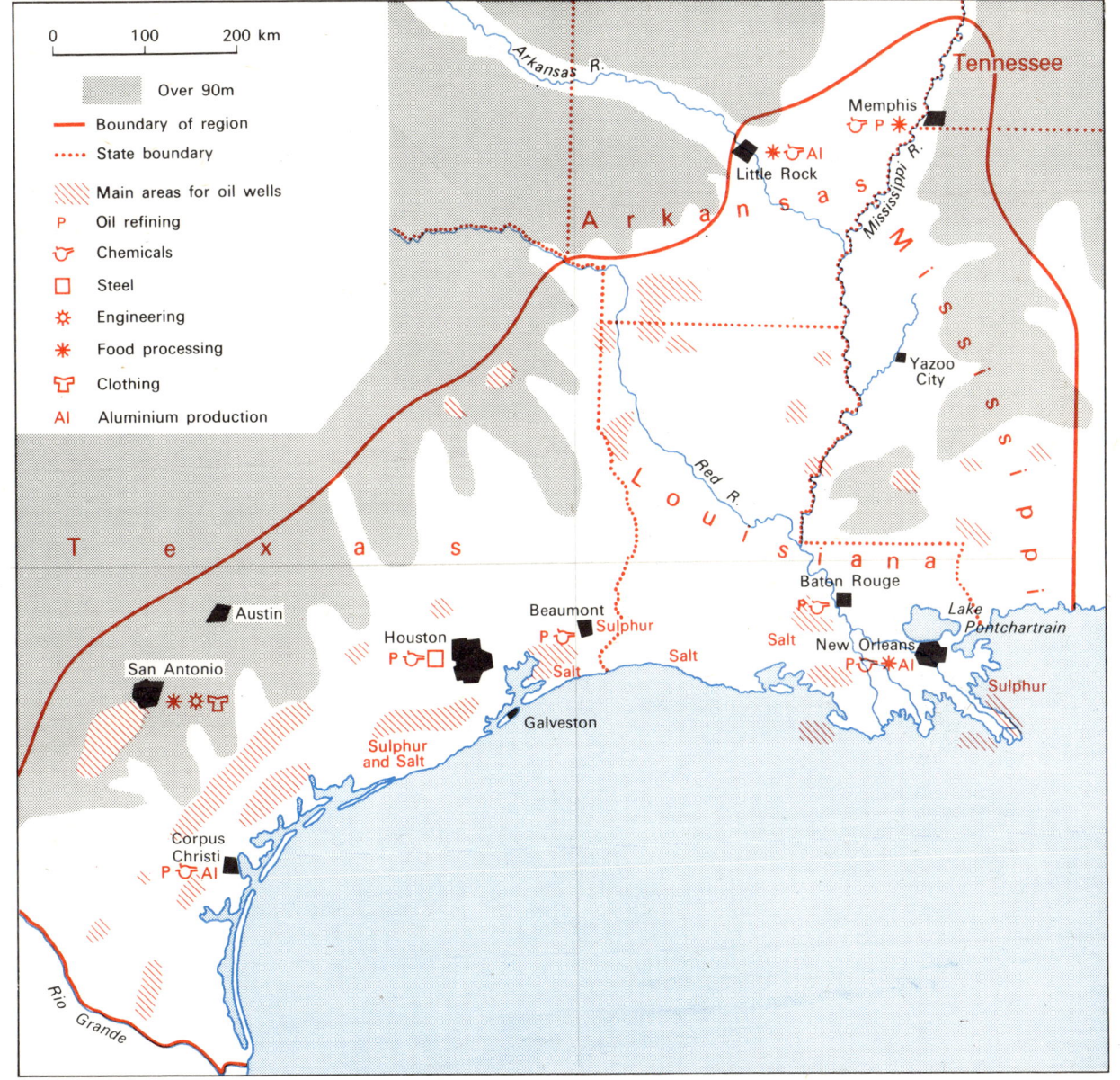

FIG. 17.4 **The Gulf Coast and Lower Mississippi – industry**

Scale:
0   100   200 km

Legend:
- Over 90m
- Boundary of region
- State boundary
- Main areas for oil wells
- P   Oil refining
- Chemicals
- Steel
- Engineering
- Food processing
- Clothing
- AI Aluminium production

Map labels:
Tennessee
Arkansas R.
Memphis — P
Little Rock — AI
Mississippi R.
Arkansas
Mississippi
Yazoo City
Red R.
Louisiana
Baton Rouge — P
Texas
Austin
Houston — P, Steel
San Antonio
Beaumont — P
Sulphur
Salt
Salt
Salt
Lake Pontchartrain
New Orleans — P, AI
Sulphur
Galveston
Sulphur and Salt
Corpus Christi — P, AI
Rio Grande

The Houston Ship Canal, Texas
The belt along the shores of the Gulf of Mexico in Texas and Louisiana has many scenes similar to this. Why?

Fairchild Aerial Surveys Inc

Mineral Working

The Gulf coast has a wide variety of minerals, including one-third of the nation's oil and natural gas as well as massive deposits of sulphur and salt. The salt forms giant domes both in Louisiana and Texas and it is on the flanks of these domes that oil is found. The oil is obtained not only from beneath the land but from offshore borings normally using mobile drilling platforms for the purpose. Sulphur occurs on top of the salt domes, and is extracted by forcing superheated water into the rocks. This melts the sulphur and the molten sulphur is either forced to the surface or can be pumped up. The major bauxite workings of the United States occur in the northern part of the region, south-west of Little Rock, Arkansas, the ore being converted into aluminium locally or in towns along the Gulf coast.

Manufacturing

The coastal belt from the delta of the Mississippi to Corpus Christi has become a major industrial region during the last 25 years. Expansion has been due primarily to the mineral resources, including oil, natural gas, salt and sulphur, but the coastal location has also been a contributory factor since water is needed as a coolant in the manufacturing processes and also provides a method of disposing of noxious effluent and a means of transport.

Here is the largest oil refining region in the United States because there is easy access not only to the local oilfield but the mid-continental one as well. It is also a major centre for the chemical industry. The sulphur is used to make sulphuric acid, which in turn is used in the manufacture of paper, explosives and fertilisers, whilst the petro-

New Orleans

U.S. Information Service

chemical industry is concerned with products ranging from plastics and synthetic rubber to nylon. The production of aluminium using Arkansas bauxite is another important industry; this in turn has attracted other companies which make aluminium products such as car parts.

Houston (938 000), which is connected to the Gulf by a deep water ship canal, is a major port dealing mainly with oil, cotton and timber. Its growth has been largely associated with the oil industry; oil refining is a major activity and it has petro-chemical industries based on the various by-products of oil.

At the other end of the oilfield is *New Orleans* (628 000). Although it lies on the Mississippi delta, 176 kilometres from the sea, there are deep water channels to the south and south-west. It is the main port for the whole Mississippi basin with 32 kilometres of wharves, its principal exports being cotton, oil and wheat. Food processing, oil refining and chemical manufacture are its chief industries.

Other large towns include *San Antonio* (588 000), *Austin* (186 000),

and *Corpus Christi* (167 000), which are general engineering and marketing centres and have food processing industries.

Away from the coast two other towns are of significance. These are *Little Rock*, the state capital and main market town of Arkansas, and *Memphis* (497 000), a river port and collecting centre for cotton, timber and agricultural produce.

Exercises

Answer in note form:
1. By means of a diagram describe the changes in land use you might expect to find on a journey from Minnesota to Louisiana.
2. What are the main features of the lower Mississippi and what has been done to prevent flooding and improve navigation?
3. Describe the site, position and function of New Orleans.
Essay:
   Write an account of the natural resources of the state of Texas and explain how these have influenced the occupations of the people.

18: THE HIGH PLAINS

The High Plains, which extend from the Canadian to the Mexican frontiers, are made up of almost horizontally bedded rocks which slope down imperceptibly eastwards from a height of about 1500 metres at the foot of the Rockies to about 500 metres on the eastern border of the region.

The plateau surface varies considerably in character according to the type of rock. In the 'badlands' of South Dakota and parts of Montana, where the rocks are mainly soft clays or unconsolidated sandstones, the land has been deeply dissected, with numerous gullies.

National Film Board of Canada
The badlands on the Canadian-American border. Describe and account for this scene.

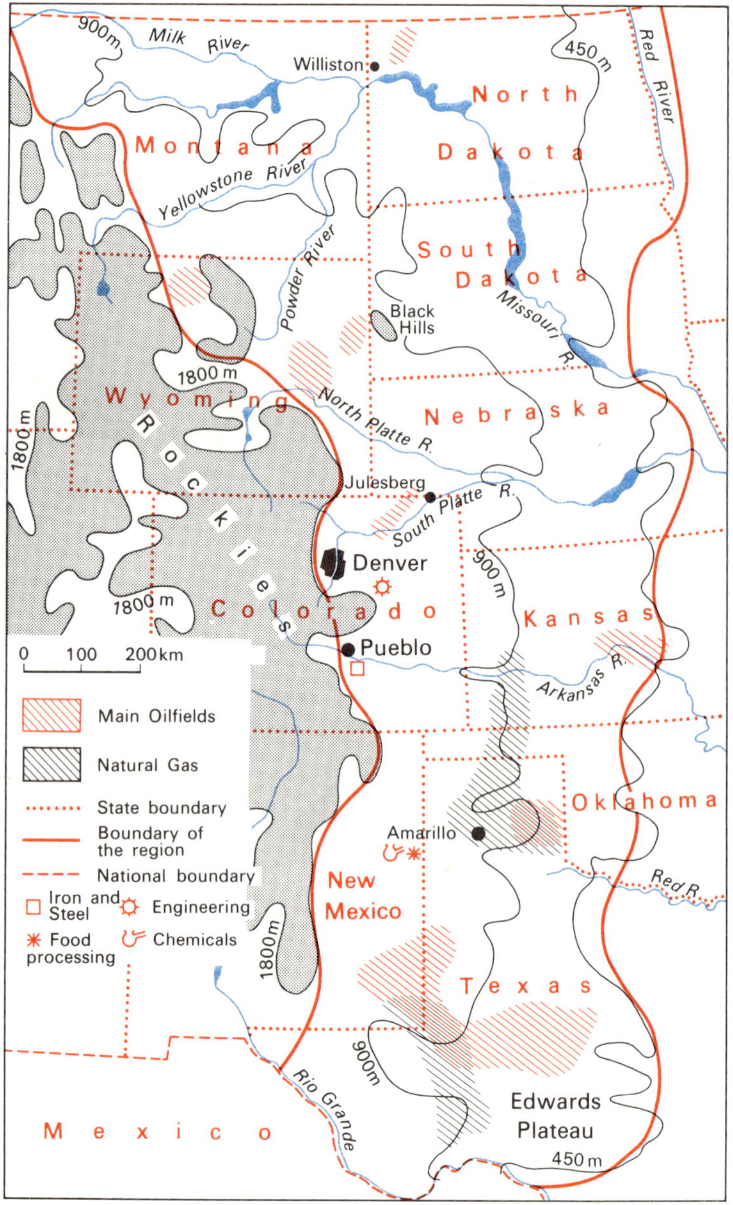

FIG. 18.1 **The High Plains**

These are the result of torrential rainstorms in an area of generally low rainfall, where there is only a sparse covering of vegetation. The Black Hills are the only prominent feature and have a core of granite which has been exposed by denudation.

Due to the great distance from the sea, the climate of the High Plains is of continental type which is characterised by great extremes of temperature and a relatively low annual rainfall, the amount decreasing from about 600 millimetres in the east to under 250 millimetres in the west. Moreover, the rainfall is unreliable and drought years have brought ruin to many farmers in the past. In summer low pressure systems can produce strong convectional currents, resulting in occasional violent storms, which is another factor unfavourable to agriculture.

Soil Erosion

Farmers have in the past been made bankrupt because of these climatic hazards and also due to wind erosion of the top soil when grasslands have been ploughed up indiscriminately for arable farming. In this way a 'dust bowl' was created over large tracts in Texas, Oklahoma, Kansas and Colorado, the most critical year being 1934 when blizzards following a long drought removed millions of tons of soil over a short period.

Today the problem is being tackled by a series of measures. The proportion of arable land has been reduced, especially in the drier

U.S. Information Service
The results of a dust storm. How can this be combated?

parts, and windbreaks of trees have been planted. As in the Tennessee Valley, strip cultivation has been introduced so that only limited strips of land are left without vegetation cover at any one time, and contour ploughing, which reduces the risk of gulleying after heavy rain, has been generally adopted. The use of irrigation has transformed many of the drier areas in recent years.

Farming

Wheat is grown in two distinct areas. There is a winter wheat belt, concentrated on Kansas, Nebraska, western Oklahoma and north-west

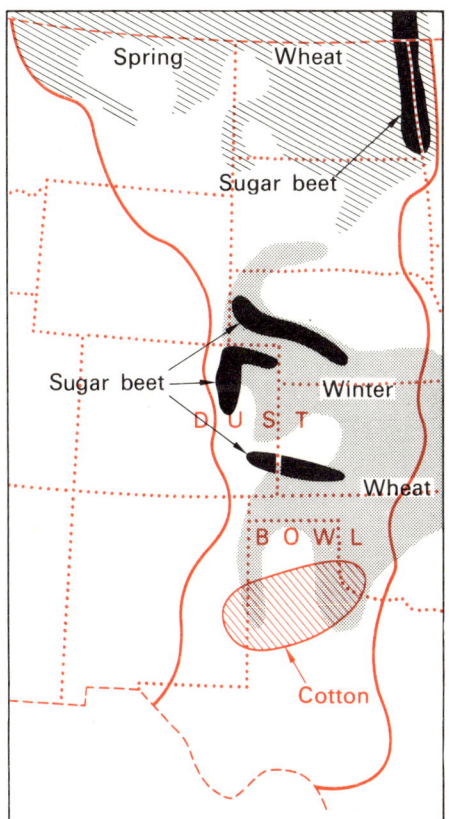

FIG. 18.2 **Croplands of the High Plains**

U.S. Information Service

Water drawn from one of nearly 200 reservoirs in Oklahoma is used on a wheat farm.

Texas, where conditions are sufficiently mild in winter to allow the wheat to survive. It is a hard, red variety which is sown in September or October and harvested in May. The spring wheat belt is in the north and is a continuation of the wheatlands of the Canadian Prairies. It is found in the Dakotas and adjoining portions of Minnesota, Nebraska, Wyoming and much of Montana, and here the wheat is sown in late April and harvested in August.

These are sparsely populated lands and the rural population has declined. As a result of greater mechanisation today a farmer can handle his 200 hectare farm without assistance except during harvest time. These grain lands are not continuous but generally occur in patches with pasture land between. Other crops besides wheat can be grown, including maize and sorghums in the south, and maize, barley, potatoes and oats in the north.

Livestock are important throughout the region. The main grazing lands are in areas where arable farming is unreliable without irrigation, i.e. where the rainfall is less than 370 millimetres per annum. Beef cattle are the main livestock and ranches vary enormously in size, ranging from a few hundred to hundreds of thousands of hectares.

Sheep are often kept on the same holding but generally graze the shorter, tougher grasses of the more arid regions. The Edwards Plateau in Texas is especially important for sheep and angora goats, and mohair from the latter is used in clothing and upholstery. Among the problems facing the farmer are parched pastures resulting from periodic droughts, especially in the south, and winter blizzards in the north. Much harm has also been done by over-grazing and some areas have been rendered useless.

Cotton growing has recently spread to western Texas where cultivation is highly mechanised and heavy yields are ensured by the use of fertilisers. In some areas the crop is grown under irrigation.

Irrigation is practised in pockets throughout the drier areas, water being obtained from wells, rivers and lakes. Reservoirs along the North Platte River supply water to irrigated areas in west Nebraska and east Wyoming, and others including the Garrison and Oahe reservoirs are to be found along the Missouri Valley. Irrigated land must give high yields to pay for the cost of irrigation, and sugar beet is profitable as well as market garden crops such as vegetables and tomatoes. However, crops of wheat, maize and alfalfa, the latter being used as hay and silage for cattle and sheep, are also grown.

One of the richest farm belts of the region is the Red River Valley on the North Dakota–Minnesota boundary. Here there are fertile, deep soils rich in humus, which make it a leading area for sugar beet and potatoes, although wheat is still the major crop.

Mineral Working

The region has considerable deposits of oil, natural gas and coal. Figure 18.1 shows the main oilfields, which include in the northern portion the Williston Basin, Powder River Basin and the Denver–Julesburg area. These have been relatively slow to develop since they are less accessible than other fields to the south. The large Texan field has about one-fifth of America's reserves and much of its development has taken place very recently. The west Texas panhandle field is especially important for natural gas. Helium, which is present to the extent of 1 to 8 per cent of the natural gas, is obtained from this field and is used in welding and in hospitals.

Although coal occurs in large quantities, much of it is of poor quality and includes lignite. It is used locally, particularly in electricity power stations, as in North Dakota and eastern Montana. The Pecos Valley is the major producer of potash in the United States, and there is a considerable gold production from the Black Hills.

Farming in North Dakota
Describe the scene including the function of the buildings and the tree belts.

U.S. Information Service

Manufacturing and Towns

The manufacturing industries of the region are scattered and suffer from isolation from large centres of population. Petro-chemical industries have recently developed, especially on the Texan panhandle oilfield, whilst food processing including sugar beet factories is mainly associated with the irrigated areas. The mineral resources of the Rockies have led to some refining of non-ferrous metals in Montana, and Pueblo in Colorado has a small iron and steel industry of local importance.

By far the largest town of the region is *Denver*, Colorado (494 000), which is an important route focus at the foot of the Rockies and a commercial centre for oil and mining companies. *Amarillo* (137 000) is the industrial centre of the Texas panhandle area, marketing meat and grain and manufacturing chemicals.

Exercises

Answer in note form :

1. Under what circumstances can soil erosion occur, and how can it be avoided ?
2. Locate the winter wheat and spring wheat areas on the High Plains and explain why they are in different areas.
3. Describe the character of pastoral farming in the region.

Essay :

Outline the distribution and significance of the mineral deposits of the High Plains, with the aid of a large sketch map.

19: THE MOUNTAIN STATES

The Western Cordillera attain their greatest width in the United States, where they form a complex system of fold mountain ranges, plateaus and basins. This chapter deals with the central and eastern mountain regions, and includes (a) the Rocky Mountains, (b) the Columbia–Snake Lava Plateau, (c) the Great Basin, and (d) the Colorado Plateau and Gila Basin.

Because of the vast size of the region and the differences of altitude, there are considerable climatic variations, but everywhere there is a marked range of temperature between summer and winter. Large areas are deficient in rainfall and the inter-montane basins, which are shielded from maritime influences by the coastal mountains, receive only about 250 millimetres per annum. The wettest areas are on the western slopes of the Rockies which receive about 750 millimetres. Towards the south the climate becomes hotter and drier.

Scrub and coarse grass covers much of the inter-montane plateau country. In the north it consists of sage brush and bunch grass, which grows in distinct tufts. Farther south is the drought-resisting mesquite and creosote bush, together with cactus and the tree-like yucca. Junipers and small pines grow on the higher ground, with true coniferous forest on the Rockies.

Until the middle of the nineteenth century this rugged, inhospitable region deterred all but the hardiest settlers, but the discovery of gold in 1859–60 brought a rush of prospectors. Many of the early workings were rapidly exhausted and the population moved away, but mining remains one of the leading occupations, especially for copper, silver, lead and zinc. Other occupations are cattle and sheep ranching, and the cultivation of irrigated tracts in some of the valleys. But the opportunities for employment are limited and on the whole the mountain states are very sparsely populated.

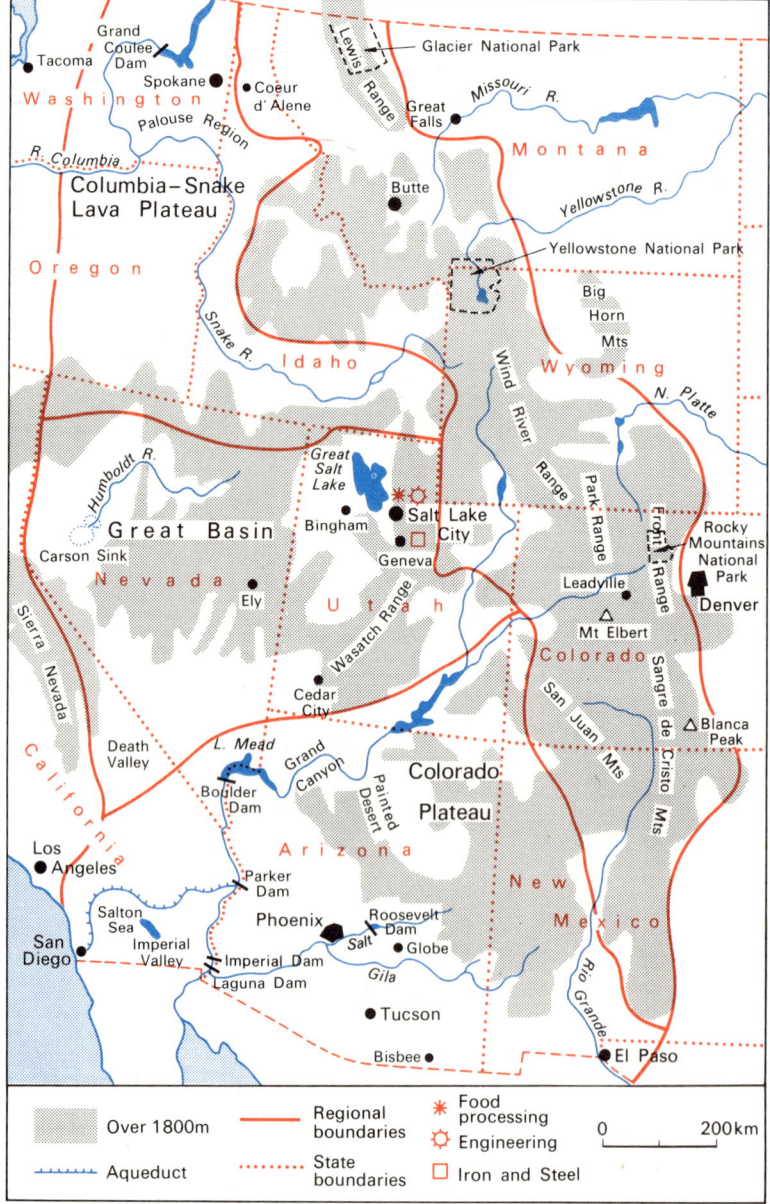

FIG. 19.1 **The Mountain States**

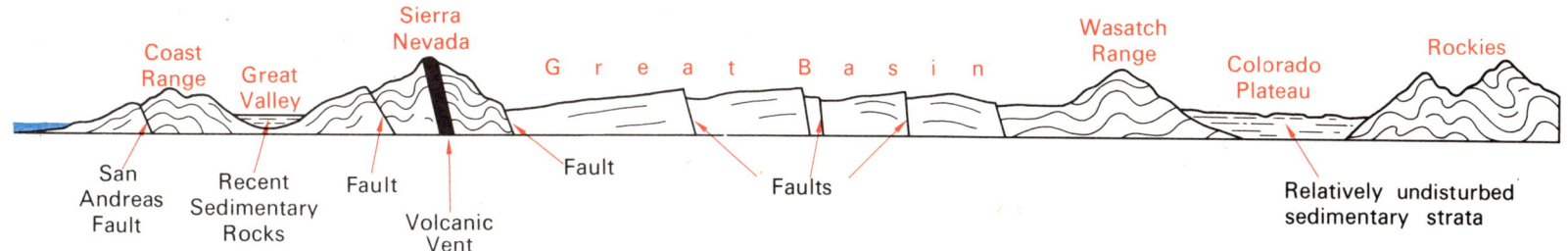

FIG. 19.2 **Section from west to east across the Mountain States region**

The Rocky Mountains

The Rockies are a zone of intense folding, with a core of ancient crystalline rocks flanked by younger sedimentary rocks. They rise steeply from the High Plains and consist of numerous ranges including the Lewis Range in Montana, the Big Horn and Wind River Mountains in Wyoming, the Front Range and San Juan Mountains in Colorado, the Wasatch Range in Utah and the Sangre de Cristo Mountains in Colorado and New Mexico. Small snowfields and glaciers surround the highest summits, some of which exceed 4250 metres above sea level, and there is widespread evidence of past glaciation in the narrow ridges, cirques, hanging valleys and deep U-shaped valleys. The lower slopes are covered by coniferous forest, with a narrow belt of mountain pasture above the tree line.

Mineral working is the most important occupation. Alluvial working has now practically ceased having been replaced by deep mining, carried on by large mining companies. Three of the main centres are *Butte*, Montana, where copper, zinc and manganese ores are worked, *Coeur d'Alene*, Idaho, which produces lead, silver and zinc, and *Leadville*, Colorado, where zinc and molybdenum are mined. Many of these ores are concentrated locally and then sent to Butte, Great Falls, Tacoma and other towns for refining.

Sheep and cattle are reared in some areas, many of the animals spending the winter in the valleys and being driven up to the high pastures for the summer. There is a little crop growing in the valleys where irrigation water can be obtained from mountain streams. Lumbering has not developed into a major industry because of the difficulty of access and distance from the coast.

The tourist attractions of the region are outstanding. In order to ensure the protection of the scenery and wild life, areas of special interest have been made into National Parks, for example the Glacier National Park in Montana, the Rocky Mountain Park in Colorado, and the Yellowstone National Park in Wyoming, with its extinct volcanoes, hot springs and geysers.

Although the Rockies present a formidable barrier to communications, they are crossed by a number of trans-continental railways on their way to the Pacific ports of Seattle, Portland, San Francisco and Los Angeles.

The Columbia–Snake Lava Plateau

Between the Rockies and Cascades, in eastern Washington, eastern Oregon and Idaho, lies a vast lava plateau which was formed as a result of repeated outpourings of molten basalt through giant fissures. Later the plateau was deeply dissected by the Columbia, Snake and their tributaries.

The basalt soils on the plateau are fertile but generally too dry for agriculture, and are used only as rough grazing land for cattle and sheep, but an exception is the Palouse region of eastern Washington where the land is higher and receives sufficient rain for wheat growing. The valley floors are intensively farmed where irrigation is practicable, the main crops being vegetables, sugar beet and alfalfa, the latter for feeding to dairy cattle. Large areas are also under orchards producing apples, pears, cherries, apricots and peaches.

Reference has already been made (page 56) to the Columbia River Project, aimed at regulating the flow of the river as well as providing water for hydro-electric power and irrigation. The potential power of the Columbia Basin is greater than that of any other river system in the United States. Nine dams have already been completed on the Columbia and a number on the Snake, each with power plant, navigation locks and fish ladders. The Grand Coulee Dam, the largest on the Columbia, was built as early as 1940. It lies across the end of a deep gorge and holds up a 240 kilometre-long lake. The water is raised by pumps and used to irrigate a fertile plain to the south of the river, and there are two large hydro-electric stations.

Most of the electricity generated in the region is transmitted to towns on or near the coast, but some is consumed locally, for example in the aluminium refining industries at Spokane, and in the huge nuclear fuel plant at Hanford, where uranium is converted into plutonium for use in nuclear weapons. The only large town in the Columbia-Snake region is *Spokane* (182 000), which is a commercial and industrial centre lying at a meeting point of transcontinental railways.

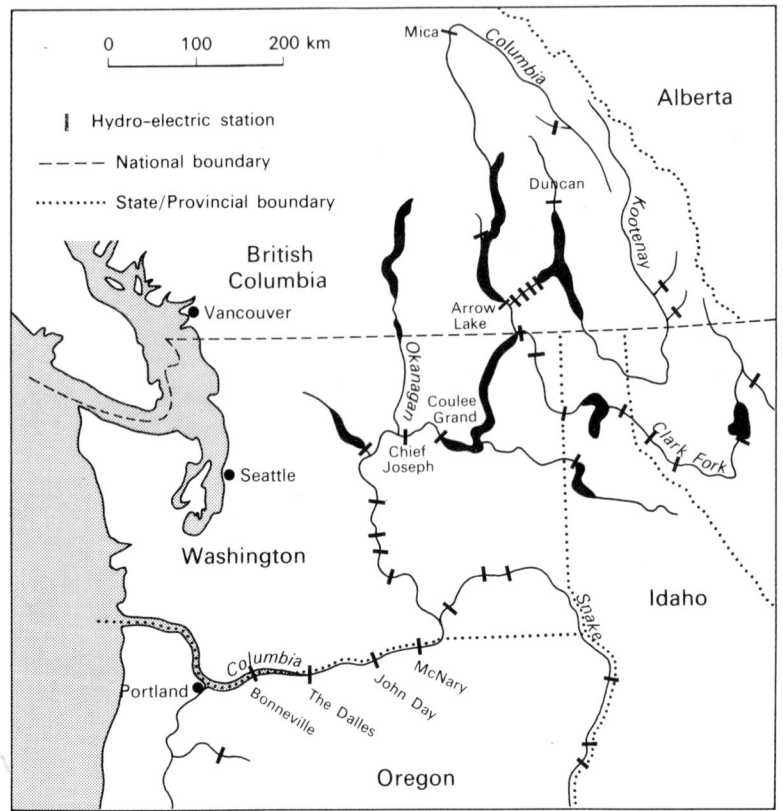

FIG 19.3 **Hydro-electric stations in the Columbia-Snake basin.**

The Great Basin

Between the Wasatch Mountains and the Sierra Nevada, mainly in Utah and Nevada, lies another large plateau area known as the Great Basin. Faulting here has produced numerous north-south ranges and depressions, including the deep rift of Death Valley, whose floor is 85 metres below sea level. Elsewhere there are wide expanses of almost level alluvium, laid down in lakes which formed towards the end of the Ice Age.

The rainfall here is as low as 250 millimetres per annum, and many streams are intermittent, flowing into basins of inland drainage and ending in shallow salt lakes or salt flats (playas). The longest of these streams is the Humboldt, which flows west to disappear into a number of 'sinks' near the foot of the Sierra Nevada. The Great Salt Lake is a remnant of a much larger Ice Age lake and is extremely shallow, having a maximum depth of only 5.5 metres.

U.S. Information Service
Describe this scene in Nevada and explain in what ways it reflects the climate of the area.

Farming is restricted to a few irrigated areas, the most important being at the foot of the Wasatch Range where streams are tapped as they leave the mountains. The chief crops are wheat, sugar beet, vegetables, fruit and alfalfa, the latter providing winter feed for cattle and sheep which graze on the hills in summer. Recently the supply of water for irrigation has been increased by the construction of tunnels through the mountains to bring water from the upper Colorado Valley.

The Great Basin contains considerable mineral wealth. Large deposits of low grade copper ore, together with some gold and silver, are mined in Bingham Canyon near Salt Lake City, partly by open cast methods, and copper, lead, silver, tungsten and zinc are worked at Ely in Nevada. In addition, over half the iron ore mined in the West comes from an area 112 kilometres west of Cedar City.

Salt Lake City (190 000) is the largest town of the region and lies to the west of the Wasatch Range, adjacent to a rich, irrigated farming area. It was first settled by the Mormons (Latter Day Saints) who came here in 1840 to escape from religious persecution. Its importance today lies in its position on a trans-continental railway route and in its manufacturing industries, including sugar refining, vegetable canning, meat packing and metal working. There are large steelworks at *Geneva* on Lake Utah, to the south of Salt Lake City, using coal from the Colorado Plateau, iron ore from Cedar City, and local limestone.

The Colorado Plateau and Gila Basin

The Colorado Plateau, which covers much of Arizona and extends north between the Wasatch Range and Rockies, is built of hori-zontally-bedded sedimentary rocks. Erosion in an arid climate has produced a tabular relief with flat-topped mesas and upstanding buttes, and the rivers have cut deep canyons. The greatest of these is the 1.6 km-deep, 200 km-long Grand Canyon of the Colorado, which has been cut in a plateau about 2500 metres above sea level and whose sides form a series of steps due to variations in the hardness of the rocks. Much of the land carries a scanty grass cover, but some areas are entirely without vegetation, as in the Painted Desert with its colourful rocks. Downstream the Colorado opens out to form the Sonora Desert, north of which lies the Imperial Valley containing the Salton Sea, 73 metres below sea level.

The giant Boulder (Hoover) Dam was completed on the Colorado in 1936. Lake Mead, which has formed above the dam, is able to store two years' normal flow of the river, so that fluctuations in

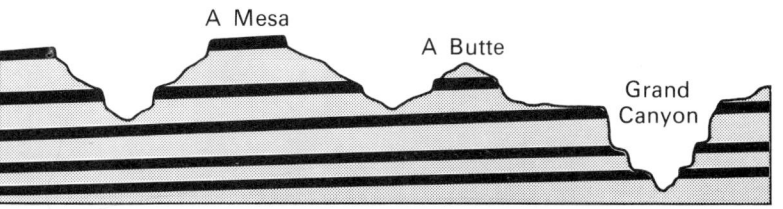

FIG. 19.4 **Section to illustrate the relief features typical of the arid Colorado Plateau**
More resistant bands of rock in solid black.

U.S. Information Service
Part of the Great Goose Necks of the San Juan River. Describe and account for the features illustrated.

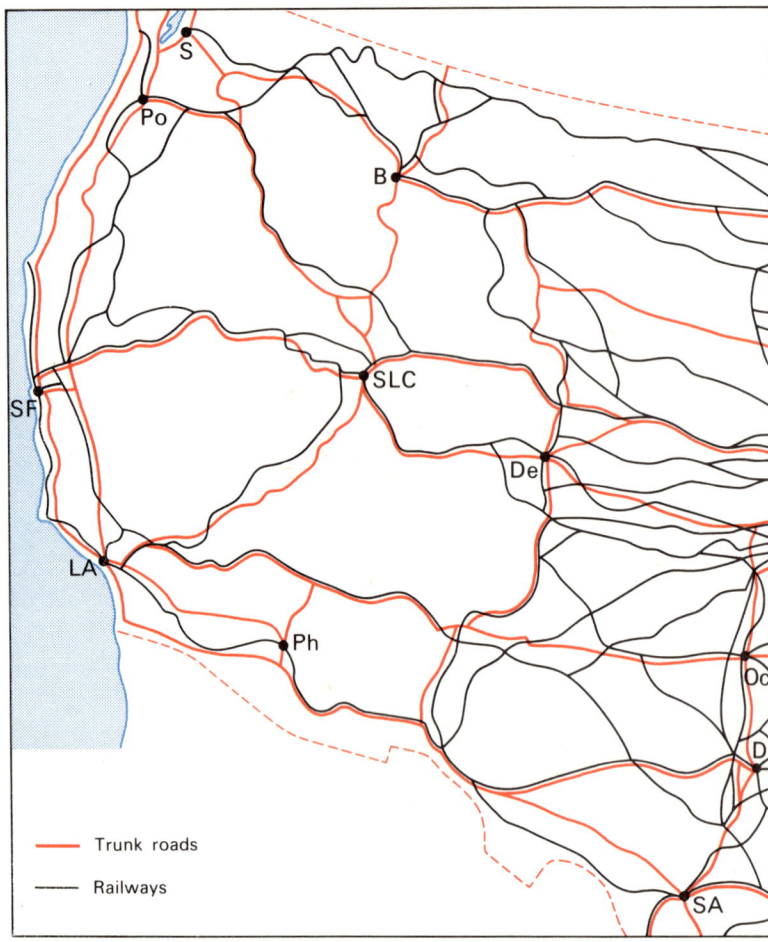

FIG 19.5 **Communications in western United States.** What conclusions can you draw from a study of this map?

U.S. Information Service

Irrigation has created a number of such rich farming areas as this in Arizona, producing cotton, citrus fruits and vegetables. ▶

volume have now been controlled. Large quantities of hydro-electricity, including half the power needs of Los Angeles, are generated at the dam.

Other dams have been built farther downstream for the supply of water and electricity. From the Parker Dam, an aqueduct nearly 1000 kilometres long carries water to Los Angeles and San Diego, and the Imperial and Laguna Dams supply water to the arid soils of the Imperial Valley.

Today the Imperial Valley is the largest irrigated area in the whole of North America, and produces high grade cotton, vegetables, citrus fruits, dates and alfalfa (for feeding to cattle and calves). Large scale irrigation is also carried on in the Salt River valley, where the Roosevelt Dam provides water for a huge area round Phoenix, and in the Gila

The Boulder Dam which created Lake Mead, 185 kms long and a recreational
area for the two and a half million visitors to the area each year.

U.S. Information Service
An open copper pit near Tucson, the centre of Arizona's copper production.

Valley. As a result of these schemes, Arizona is now the leading cotton-producing state in the U.S.A.

There is mineral working at a number of places. Half the United States' output of copper comes from Arizona, where the ore is dug from huge open pits near Bisbee and Globe. Zinc, lead, coal and iron ore are also mined. In the far south live descendants of the early Indian inhabitants, notably the Hopi and Navajo, in pueblos (villages) of flat-roofed houses made of stone or sun-dried mud. They grow maize and keep livestock, and to augment their income they produce wood carvings for sale to tourists.

Exercises

Answer in note form:

1. What do you understand by lava plateau, playa, canyon? Give examples of each.
2. Describe the natural vegetation of (a) the Rockies and (b) the inter-montane plateaus.
3. Make a list of the most important minerals mentioned in this chapter and state where they are worked.

Essay:

What part do the rivers play in the economy of the Mountain States?

20: THE NORTH WEST

This region covers the western parts of the states of Washington and Oregon, which border the Pacific Ocean. In spite of its valuable timber resources, rich fishing grounds and fertile lowlands, it remained undeveloped until the nineteenth century, mainly because of its remoteness and isolation from the eastern states. Settlers began to arrive after 1840, but development was slow until the completion of the trans-continental railways in the 1890s. During the present century there has been a rapid increase in population and all branches of the economy, including lumbering, fishing, farming and manufacturing, have shown a remarkable expansion.

Physically the region consists of three north to south belts:

(1) In the interior is the lofty Cascade Range which rises to a number of volcanic cones exceeding 3600 metres above sea level, including Mount Rainier (4392 metres).

(2) To the west of the Cascades is a central depression stretching from the Puget Sound lowlands to the Cowlitz and Willamette valleys. These lowlands have generally fertile soils, formed on alluvial or glacial deposits.

(3) Overlooking the Pacific is the Coast Range, consisting of more rounded hills which seldom exceed 1200 metres above sea level, although Mount Olympus in the north reaches 2404 metres. The Columbia River has cut a deep trench through both the Cascades and Coast Range to reach the Pacific.

The climate of the North West, like that of British Columbia, is of cool temperate western margin type, since the region is under the influence of south-westerly winds from the Pacific Ocean. Winters are mild and summers fairly cool, and there is precipitation throughout the year, derived mainly from depressions. The amount varies from over 2000 millimetres per annum on the exposed windward slopes of the Coast Range and Cascades, to only 750 millimetres in the sheltered central depression. Areas to the east of the Cascades lie in a marked rain shadow and have as little as 250 millimetres.

The humid atmosphere, heavy precipitation and long growing season have produced valuable coniferous forests on the mountain slopes. The most important tree is the tall, straight Douglas Fir, and others are hemlock, cedar, spruce and, at higher altitudes, pine.

Forest Industries

The Coast Range and Cascades, together with the Columbia Basin, are America's main source of softwood timber. The Douglas Fir supplies two-thirds of the total production and is in great demand for constructional work. Most of the timber is sent by road, rail or sea to other parts of the country, but some is used in local woodworking industries for the manufacture of plywood, veneers, hardboard and furniture. The production of pulp and paper is relatively unimportant.

Lumbering presents considerable difficulties because of the rugged, mountainous nature of the land. Felling and transporting are fully mechanised, and specialised devices are employed including the use of overhead cables for dragging the logs down steep slopes. Many of the sawmills and pulpmills are situated on tidal water, chiefly on Puget Sound and along the Columbia River, but some lie in the interior lowlands, where they are served by roads and railways. All have access to hydro-electric power.

Fishing

Measured by value, the fisheries of Washington and Oregon are the most important in the United States. By far the main catch is salmon, a considerable proportion of which is canned. Tuna, halibut, herring and pilchards are caught in deeper water, and oysters are farmed in some of the bays. The public authorities are taking strong measures to conserve the salmon fisheries by building fish ladders at the hydro-electric dams, by operating hatcheries from which the young fry are taken to the rivers, and by preventing river pollution. Seattle has a local fishing fleet and is also the main base for the Alaskan fleet. The greatest salmon river is the Columbia, and near its mouth is the important fishing port of Astoria.

Farming

The best farmlands lie on the fertile lowlands round Puget Sound, in the Cowlitz and Willamette valleys, and in a number of smaller valleys near the coast. The cool temperate climate encouraged the first farmers to grow wheat as the main crop, but competition from the farmers of the Great Plains made this unprofitable, and the emphasis

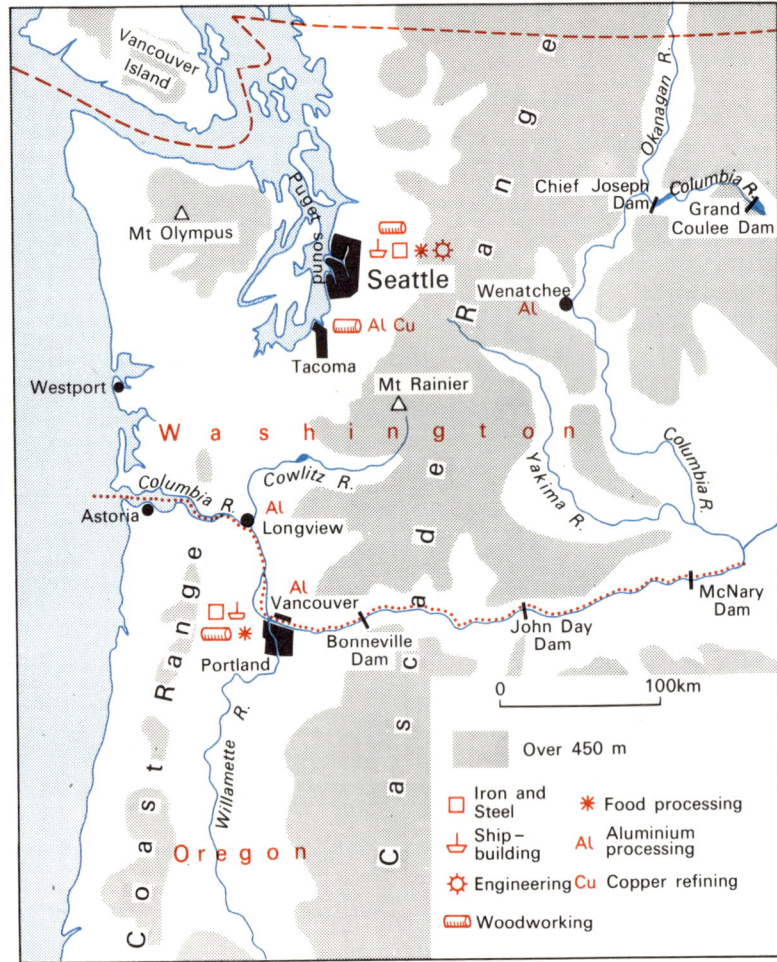

FIG. 20.1 **The North West**

Map legend:
- □ Iron and Steel
- ⚓ Ship-building
- ☼ Engineering
- ▭ Woodworking
- ✳ Food processing
- Al Aluminium processing
- Cu Copper refining
- Over 450 m

The White Salmon River at its junction with the Columbia.

has changed to dairying, poultry keeping and truck farming, mainly to satisfy the demand in local urban areas. Another activity is the growing of deciduous fruit, including apples, pears, peaches, plums, cherries and bush fruits, and also flowers. In the drier areas there is irrigation in summer, chiefly from overhead sprinklers.

Manufacturing and Towns

The first industries to develop in the region were concerned with the processing of local timber, fish and farm produce, and these are still of great importance. A serious disadvantage is the lack of useful minerals, and essential supplies of coal, oil and natural gas have to be brought in from other parts of the United States and from Canada.

However, there is an abundance of hydro-electric power. The heavy, reliable rainfall, deep valleys and melting snow in the mountains provide excellent conditions for hydro-electric developments, and it has been estimated that about one-quarter of the country's hydro-electric potential lies in the North West, including the Columbia Basin. There are numerous power stations in the Willamette depression, in the smaller river valleys and in the Puget Sound area, but these are dwarfed by the latest power developments in the Columbia Basin, details of which were given in the previous chapter.

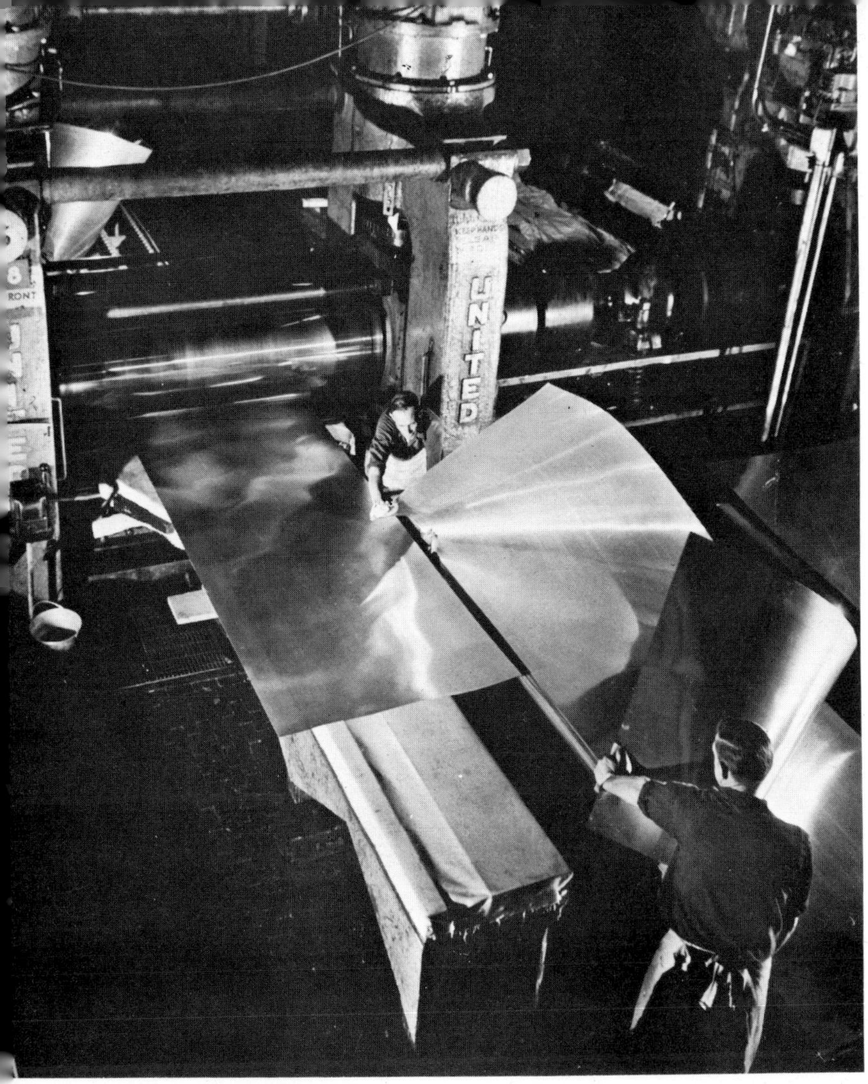

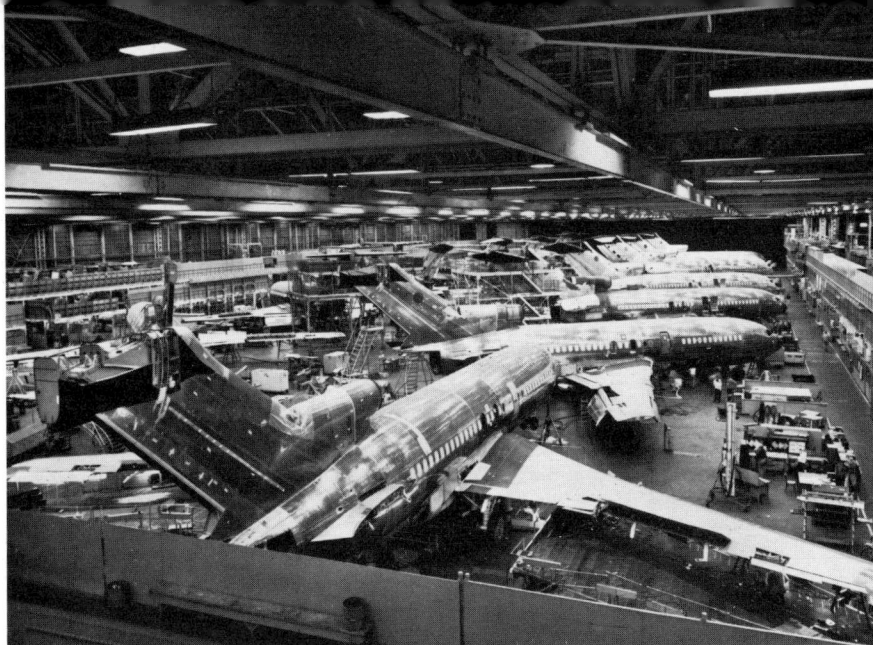

Aluminium rolling *U.S. Information Service*

U.S. Information Service
The Boeing plant at Renton, near Seattle. Why is there such a plant in the North West?

One of the most important uses of this cheap electricity is the production of aluminium. Bauxite, obtained from Jamaica and the Guianas, is first processed on the Gulf coast, and the alumina so obtained is railed to the North West, including the Portland and Tacoma areas, for the final refining process. The smelting and refining of copper at Tacoma, using ores from the Rockies, is also based on hydro-electricity. Very little steel is produced in the region, but metal-using industries such as shipbuilding, motor vehicle and aircraft manufacture have developed in a number of towns.

Seattle (565 000) is the largest town and leading port of the region, being situated on the east side of the deep, sheltered Puget Sound. It is the terminus of trans-continental railways and has a large commercial trade especially with Alaska and the Far East. It is a fishing port and its industries include fish processing and canning, timber working, shipbuilding, car and aircraft manufacture, and a small steel industry based on scrap metal. Farther south on Puget Sound is the smaller port of *Tacoma*, among whose industries are copper and aluminium refining.

Portland (382 000) lies at the confluence of the Columbia and Willamette Rivers, about 160 kilometres from the sea. It occupies a nodal position, being at the crossing place of the north-south route following the Willamette and Cowlitz Valleys and the east-west route along the Columbia Valley. The lower Columbia has been dredged to allow sea-going ships to reach the city, and upstream the river is navigable as far as Idaho. Thus Portland has developed into a market and commercial centre serving a vast and productive hinterland. Its principal industries are food processing, timber working, shipbuilding and aluminium refining.

Exercises

Answer in note form :

1. Why is forestry an important occupation in the North-West, and what uses are made of the trees?
2. What types of farming are carried on in the lowlands?
3. Compare the climate of the North-West with that of the High Plains.
4. Draw an annotated section to show the physical features you would see on a journey from Spokane to Seattle.

Essay :

Describe and give reasons for the distribution of population in Washington and Oregon.

U.S. Information Service Photographs ▶

The top photograph shows American Indians fishing for salmon with hoop nets in Oregon. They have perpetual treaty rights to fish in these waters. The lower scene shows the unloading of salmon at Astoria where they will be canned. Fishing is done from diesel powered boats which work from the mouth of the Columbia River to 80 kilometres offshore.

21: CALIFORNIA

California was settled by the Spaniards in the eighteenth century, and even today Spanish influence is apparent in many place names and in the architecture of churches, public buildings as well as some private homes.

The main influx of settlers from the eastern United States dates from the discovery of gold in the Sierra Nevada in 1849. Most of the alluvial workings were rapidly exhausted, but some of the miners stayed on to farm the land. Irrigation schemes opened up vast new areas, and the completion of the trans-continental railways made it possible to market the farm produce in the eastern states. Migrants poured into the region, many from the exhausted farmlands of the High Plains.

During the present century people have continued to move west in their thousands, attracted by the fine climate and opportunities for employment in the rapidly growing industries. Today California's population is more than 18 million, and it has overtaken New York to become the most populous state of the Union.

The physical features of California are a continuation of those of Washington and Oregon, consisting of eastern and western mountain ranges and a central depression.

(1) The eastern range, the Sierra Nevada, contains some of the highest mountains in North America, including Mount Whitney, 4418 metres. It is an uplifted massif of ancient rocks into which igneous rocks have been intruded, and glaciation has left its imprint in the sharp-edged summits and deep cirques.

(2) The central depression is known as the Great Valley of California, and is a level tract about 80 kilometres wide and 640 kilometres from north to south. Much of its floor is covered by fertile alluvium and it is drained by the Sacramento and San Joaquin rivers, which unite to enter San Francisco harbour in a swampy delta.

(3) The Coast Range is much lower than the Sierra Nevada and is the result of both folding and faulting. It ends in high cliffs and headlands along the coast, with a large break at the drowned inlet of the Golden Gate and San Francisco harbour. The occurrence of earthquakes shows that earth movements have not entirely ceased. The main line of weakness is the San Andreas fault: movement along this has been the cause of most of the earthquakes including that which largely destroyed San Francisco in 1906.

Most of California has a Mediterranean type of climate, with hot, dry summers and mild, rainy winters. Summer temperatures are less high along the coast because of the influence of the cool California Current, and the mixing of cold and warm air produces persistent fog in the San Francisco area. Conditions become more extreme in the Great Valley, where summers are hotter and there are night frosts in winter.

The annual precipitation decreases from about 500 millimetres at San Francisco to under 250 millimetres in the southern part of the Great Valley, but is as high as 1000 millimetres on the western slopes of the Coast Range and Sierra Nevada. To the south and south-east the hot, dry season lengthens and the cool, wet season becomes shorter, until the desert is reached.

The natural vegetation, as in other Mediterranean regions, consists of evergreen trees and shrubs, many having small, leathery leaves and long roots which permit them to withstand the long, summer drought. This kind of vegetation is known in California as chaparral. On the mountains, where the rainfall is heavier, are coniferous forests including giant redwoods and Douglas Fir in the northern part of the Coast Range, and sequoias in the Sierra Nevada. The sequoias and giant redwoods are the world's largest trees, many specimens growing to heights of nearly 100 metres and having trunks 6 to 10 metres in diameter. On the drier mountain slopes the most important tree is the ponderosa pine. Cacti and sage brush grow in the most arid areas, with salt desert shrubs in the south.

Forest Industries and Fishing

The forest wealth of the northern Coast Range and Sierra Nevada makes California the second timber-producing region in the country, after the North West. Most of the lumbering takes place on mountain slopes between 1000 and 2000 metres above sea level, and almost all the logs go to sawmills.

A wide variety of uses is made of the timber. The redwoods, which are water-resisting, are ideal for outdoor purposes such as fencing and roofing, and are also much used for furniture. The Douglas Fir is employed in general constructional work including house-building, and the softer ponderosa pine mainly for indoor purposes. Large quantities of timber are also used in making boxes for the fruit-packing industry.

Fishing is another important occupation. Although the continental shelf is narrow off the Californian coast, there is abundant plankton and a rich and varied fish life. Sardines, mackerel, pilchards and tuna are

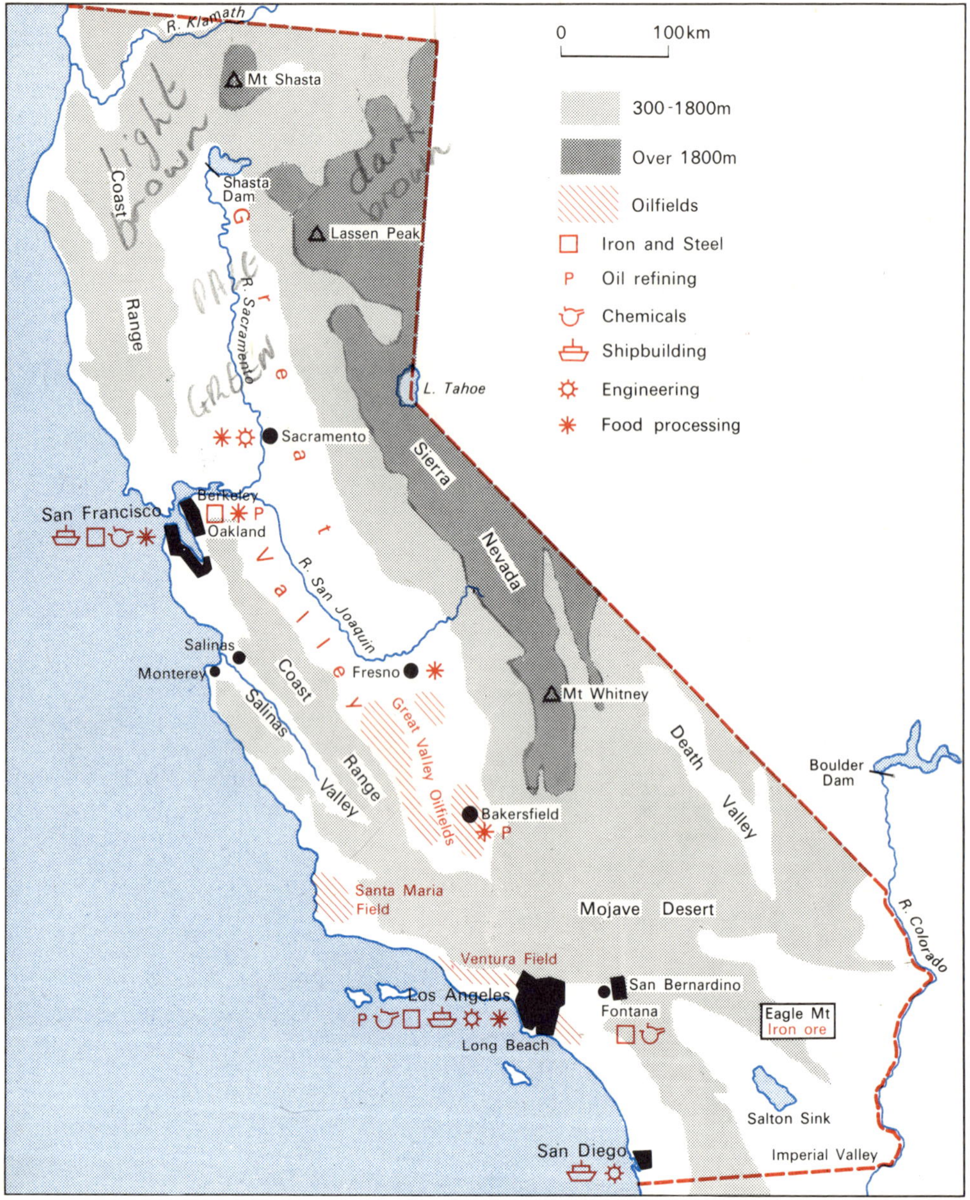

U.S. Information Service
Redwood stand in north California.

FIG. 21.1 **California**

Map labels:
R. Klamath
Mt Shasta
light brown
dark brown
Coast Range
Shasta Dam
Lassen Peak
PALE GREEN
R. Sacramento
Great Valley
L. Tahoe
Sacramento
Berkeley
San Francisco
Oakland
Sierra Nevada
R. San Joaquin
Salinas
Monterey
Fresno
Coast Range
Salinas Valley
Mt Whitney
Death Valley
Great Valley Oilfields
Bakersfield
P
Boulder Dam
Santa Maria Field
Mojave Desert
R. Colorado
Ventura Field
Los Angeles
San Bernardino
Fontana
Eagle Mt Iron ore
P
Long Beach
Salton Sink
Imperial Valley
San Diego

Legend:
300-1800m
Over 1800m
Oilfields
Iron and Steel
P Oil refining
Chemicals
Shipbuilding
Engineering
Food processing

0 100km

caught in local waters, and some vessels bring in large catches from as far away as Alaska and Hawaii. The principal fishing ports are San Francisco, Los Angeles, Monterey and San Diego.

Farming

Agriculture is of great importance both in the Central Valley and on the coastal lowlands. In the wetter northern parts of the Sacramento valley large numbers of sheep and cattle are kept, and wheat, barley and alfalfa are the chief crops, but elsewhere in the Central Valley much of the land is under fruit (peaches, apricots, plums, grapes and citrus fruits), nuts (mainly almonds and walnuts), and vegetables. In addition, rice is grown in the delta region and cotton in the southern San Joaquin valley.

Almost all this depends on irrigation, and it is fortunate that there is heavy precipitation in the mountains, some of which is stored as snow until early summer. The rivers are dammed where they leave the high ground, which serves to regulate their flow and to provide

U.S. Information Service
Irrigation flooded rice fields in the delta area of the Sacramento.

FIG. 21.2 **Irrigated areas in the Great Valley of California**

hydro-electricity, as well as water for irrigation. The most outstanding irrigation work is the 800 kilometre-long canal which carries water from above the Shasta Dam on the upper Sacramento to the parched lands of the south.

There are also a number of fertile valleys in the Coast Range, including the Salinas Valley where lettuces, carrots, beans and sugar beet are grown on a large scale. The Los Angeles lowland specialises in citrus fruit and the orchards, which are irrigated from wells and streams, are planted on sloping ground above the hollows where cold air accumulates at night. Much valuable orchard land has been lost in recent years as a result of the expansion of Los Angeles and other towns.

California contains one-quarter of the total irrigated land in the United States, and produces nearly half of the country's fruit. Large quantities of fresh fruit and early vegetables are sent to the eastern states by road, rail and even by air. Much of the produce is processed locally, including the canning and quick-freezing of fruit and vegetables, the extraction of orange juice, the drying of prunes and raisins, and the making of wine.

Minerals

There are large deposits of oil and natural gas in the southern part of the Great Valley, and in the Santa Maria, Ventura and Los Angeles lowlands, and these have played a great part in California's recent industrial development. Crude oil is piped to refineries in Los Angeles, San Francisco and other towns, and natural gas, which occurs with the oil, is also distributed by pipeline. California is the third most important oil-producing state after Texas and Louisiana, with an annual output of about 45 million tonnes. However, production is now declining and is not sufficient for local needs: about one-third of the oil used in the state has to be imported.

California is the leading gold-producing state and the main workings are still on the western side of the Sierra Nevada, where gold was first discovered in 1848. The alluvial gold is now worked by huge mechanical dredgers, and there is also deep mining in the mountains.

Limestone, for which there is a great demand for the making of cement and concrete, is worked in the Coast Range, and other valuable minerals are iron ore, tungsten and mercury.

Manufacturing and Towns

In industrial output California is now second only to New York State. Before the 1939–45 war manufacturing took the form mainly of the processing of farm produce, but during and since the war there has been a remarkable growth of new industries. As a wartime measure a large modern steelworks was erected at Fontana, 80 kilometres to the east of Los Angeles, using Utah coal and Eagle Mountain ore, and this, together with smaller steelworks at Los Angeles and San Francisco, supplies most of California's requirements.

Metal-using industries, including shipbuilding and the manufacture

of motor vehicles, aircraft and electronic equipment, have also developed in the main towns, and there are varied chemical industries, using oil and natural gas as their raw materials. The factors responsible for this rapid industrialisation can be summed up as follows:

(1) The large population of the region provides plenty of labour and a vast market for the finished goods.

(2) The heavy freight charges on goods brought in from the eastern states have encouraged the setting up of industries in California.

(3) Power supplies are plentiful, being derived almost equally from thermal power stations (using oil and natural gas) and hydro-electric stations, including that linked with the great Boulder (Hoover) Dam on the Colorado.

(4) There is considerable wealth of industrial raw materials, of which the most important are farm produce, timber, limestone and metal ores, as well as oil and natural gas.

The warm, sunny climate, magnificent mountain and desert scenery, and fine beaches attract millions of tourists each year to California, and for the same reasons many retired people have made their homes here.

Eighty-five per cent of the state's population live in towns and suburban areas, especially in the two great conurbations of San Francisco and Los Angeles.

San Francisco lies on a hilly peninsula just south of the Golden Gate, on one of the finest natural harbours in the world, and is part of a vast conurbation including Oakland, Stanford, Berkeley and other towns, whose total population exceeds 5 million. The irregular shape of the harbour made communications difficult before the completion, just before the war, of the Oakland–San Francisco and Golden Gate bridges.

San Francisco is the western terminus of an important transcontinental railway and is the natural outlet of the rich farmlands of the Great Valley. It handles most of America's trade with the Far East, Australasia and west coast of South America. The conurbation has developed a great variety of industries including steel production, shipbuilding, car assembly, oil refining, petro-chemicals and food processing, and in recent years it has become a leading centre for the manufacture of computers, space rockets and missiles.

Los Angeles is situated on an arid lowland, overlooked from the north and east by the San Gabriel and San Bernardino Mountains. It can best be described as a series of sprawling suburbs which include Glendale, Pasadena, Alhambra, Santa Monica, Hollywood

San Francisco   *U.S. Information Service*

and the ports of San Pedro and Longbeach. The conurbation is one of the largest in the world and has a total population of over 8 million.

The abundant natural resources of southern California have made Los Angeles one of the leading industrial centres in the United States. Its products range from steel, aircraft, motor vehicles and ships to computers and space equipment, and there are huge oil refineries with associated chemical industries.

The 1920s saw the growth of the motion picture industry at Hollywood, but today many of the studios have to be content with making feature and serial films for television. The clear air and sunshine which encouraged the early film industry can still be found in the surrounding hills, but most of the conurbation now suffers from persistent smog, caused by the exhaust fumes of millions of cars and industrial waste gases. There is an acute water supply problem, too;

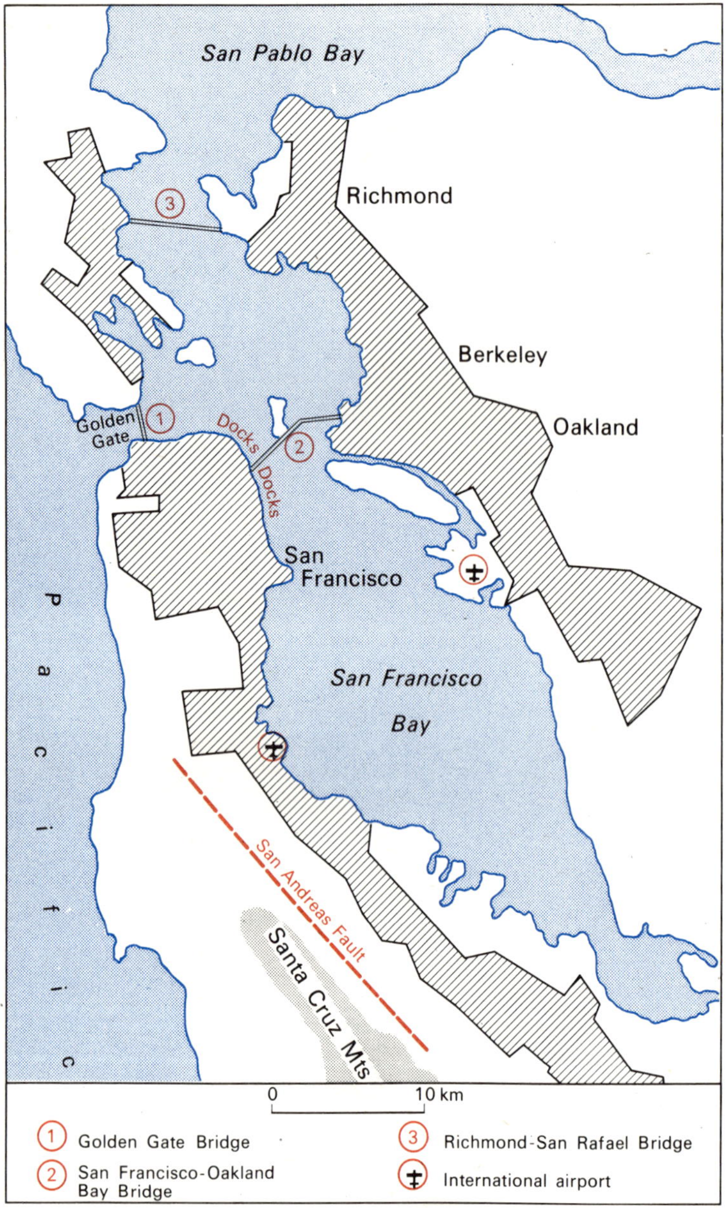

FIG. 21.3 San Francisco

San Pablo Bay

Richmond

Berkeley

Oakland

Golden Gate

Docks Docks

San Francisco

San Francisco Bay

P a c i f i c

San Andreas Fault

Santa Cruz Mts

①	Golden Gate Bridge	③	Richmond-San Rafael Bridge
②	San Francisco-Oakland Bay Bridge	✠	International airport

FIG. 21.3 **San Francisco**

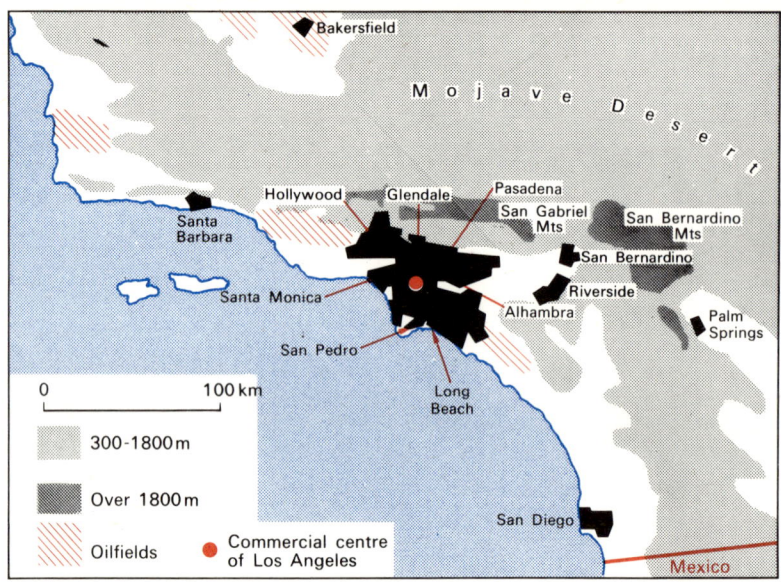

FIG. 21.4 **Los Angeles**

Bakersfield

M o j a v e   D e s e r t

Santa Barbara

Hollywood — Glendale — Pasadena
San Gabriel Mts

San Bernardino Mts

Santa Monica

San Bernardino

Riverside

San Pedro

Alhambra

Palm Springs

Long Beach

San Diego

Mexico

0	100 km

300–1800 m

Over 1800 m

Oilfields

● Commercial centre of Los Angeles

water has to be obtained from as far away as the Colorado and Sierra Nevada, and to meet future requirements it may be necessary to bring water from the Columbia Valley 1760 kilometres away, or to introduce new processes such as the desalination of sea water linked with the operation of nuclear power stations.

San Diego (573 000) lies near the Mexican border and is a fishing port and naval base. It has aircraft factories and shipyards specialising in the construction of small craft.

Exercises

Answer in note form:

1. What do you understand by a Mediterranean type of climate and why is it found in California?
2. Write an account of fruit growing in the Great Valley.
3. Compare the sites, positions and functions of San Francisco and Los Angeles.

Essay:

Locate the main manufacturing industries of California and give reasons for their recent expansion.

22: ALASKA

Alaska is a broad peninsula in the north-west of North America, reaching to within 80 kilometres of the Siberian frontier across the Bering Strait, and including a narrow 'panhandle' down the Pacific coast. In area it covers about one-fifth of the whole United States, of which it became part as recently as 1867, when it was purchased from Russia for less than two million pounds.

Structurally, Alaska is part of the Western Cordillera, and resembles British Columbia in that it consists of two highly folded mountain zones (one close to the Arctic Ocean in the north, and the other along the Pacific coast in the south), separated by a central plateau. Its climate is not as cold as many people imagine, for two-thirds of it lie south of the Arctic Circle, and most of its harbours are ice-free in winter.

FIG. 22.1 **Alaska**

Apart from a number of gold rushes between 1880 and 1900, there was little economic development in Alaska before the 1930s, but during the last war it became a base for operations against the Japanese, and a number of large airfields were constructed, as well as the Alaska Highway linking it with Edmonton in Alberta. An important step was taken in 1959 when Alaska became the 49th state of the Union, but even today its population numbers only 250 000, including about 40 000 Indians and Eskimos.

It is a land of tremendous unused natural resources, whose development awaits only the arrival of more people and the investment of capital. Oil may provide the necessary stimulus, for the newly-discovered oilfield on the 'north slope' may prove to be one of the richest in the world.

Southern Alaska

The south coast of Alaska is rugged and indented, with many steep-sided fiords, and is bordered by two almost parallel ranges of mountains. The seaward range is represented by numerous offshore islands in the Panhandle, and is continued by the St Elias Mountains, which swing round to end in the Kenai Peninsula and Kodiak Island. Farther inland the second line of mountains forms the Wrangell and Alaska Ranges and ends in the Alaska Peninsula and Aleutian Islands.

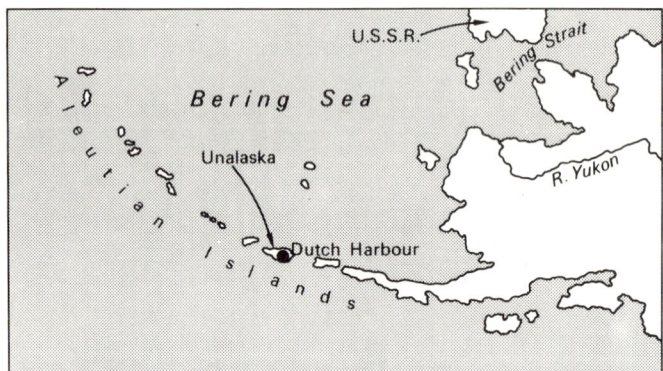

FIG. 22.2 **The Aleutian Islands**

These mountains contain the two highest peaks in North America, Mount McKinley (6193 metres) and Mount Logan (6051 metres), and are covered with extensive snowfields and glaciers, some of which descend to sea level. Among the summits are a number of active volcanoes.

The climate of southern Alaska is similar to that of western British Columbia, being under the influence of south-westerly winds which blow off the warm waters of the North Pacific Drift. Coastal areas are remarkably mild for their latitude, and have heavy precipitation. Juneau, in the Panhandle, with a January mean temperature of $-2\frac{1}{2}°C$, a July mean of $15°C$, and a precipitation of 2280 millimetres, is typical.

The lower mountain slopes are thickly forested with hemlock, spruce and cedar. Much of this valuable timber remains unused, although lumbering has developed in a few places. Some of the logs are processed locally, and there are modern pulp mills at Ketchikan and Sitka, but large quantities are also shipped to Seattle. Fur trapping is carried on in the forests, particularly for fox, muskrat and mink, and fur farming is on the increase.

The leading occupation is fishing. About half the world's catch of salmon comes from Alaskan waters, and there are numerous canneries in the coastal inlets. The salmon are caught by seiners, which trap the schools of fish within huge nets, and also by smaller boats using long lines to which are attached baited hooks. Ketchikan, with twelve canneries, is the most important salmon-fishing port. Other fish caught are herring, halibut and cod; and king crabs are canned at Homer in the Kenai Peninsula. The Pribilof Islands in the Bering Sea are the chief breeding ground for the North Pacific seal, but in order to maintain their numbers, killing has to be restricted to a short season. Dutch Harbour, on the island of Unalaska in the Aleutians, is the main base for sealing and whaling.

Another occupation is mining. Gold is obtained mainly from placer deposits, giant dredgers being used to dig up and process the alluvial material; platinum, silver and copper are also worked, although the output is small. There are a number of coal mines, especially in the Matanuska Valley north of Anchorage, and a small oilfield in the Kenai Peninsula.

There has been very little agricultural development so far, except in a few valleys including the Matanuska Valley where hardy cereals, vegetables and fodder crops are grown, and cattle, pigs and poultry are kept. The climate here is drier than elsewhere, the soils more fertile, and the long hours of daylight ensure rapid growth.

Southern Alaska has a vast hydro-electric potential, only a tiny proportion of which is exploited at present to satisfy demand in existing towns and industries. Further developments will depend on the growth of population and economic progress.

Towns in the region are generally small. The largest is *Anchorage*

Anchorage   *British Petroleum*

Juneau   *U.S. Information Service*

Using the photos, map and text, describe the sites, positions and functions of these towns.

(45 000), which lies at the head of Cook Inlet on the railway from Seward to Fairbanks. A fishing fleet operates from the port, and it has an international airport serving as a re-fuelling point for aircraft on the trans-Polar flight from Europe to Tokyo. *Juneau* (7 110) is the state capital, but its growth has been hampered by poor communications with the interior. The small port of *Skagway* is the starting point for the railway to the Klondike, in the Canadian Yukon. These and other towns are mainly commercial and market centres, exporting local produce and receiving supplies by coastal shipping services.

The Central Plateau

Central Alaska is a dissected plateau, consisting of a series of flat-topped uplands separated by broad valleys, drained by the Yukon and a number of other large rivers. The climate is of extreme continental type, with long, severe winters, rather cool summers and only light precipitation. As an example, Fairbanks has a mean January temperature of −25°C, a July mean of 17°C, and an annual precipitation of 280 millimetres.

The natural vegetation is thin coniferous forest and tundra. The tundra consists largely of mosses and lichens, which provide food for herds of reindeer introduced into the region in the 1890s for the benefit of the local Eskimo population. The reindeer flourished and increased in numbers, but the pastures were damaged by over-grazing and this has led to a reduction in the herds in recent years. Both reindeer and wild moose are useful sources of meat.

There are large areas of permafrost where the poor natural drainage makes farming difficult. Nevertheless a start has been made with the growing of vegetables and keeping of cattle in the vicinity of Fairbanks. Placer gold mining is carried on especially round Nome and Fairbanks, but it is mainly a summer occupation since running water is needed for washing the gold-bearing gravels.

The region is somewhat difficult of access, but can be reached by rail from Seward and also via the Alaska Highway. The Yukon River is navigable in summer, but its shallow delta limits the size of vessel that can use it. Because of the freezing of the Bering Sea, the port of Nome is open only for three months each year. The main town is

The Brooks Range

British Petroleum

British Petroleum

Put River No. 1 camp, North Slope
What problems face the oilmen in North Alaska and how might these problems be overcome?

Fairbanks (13 500), which is a route centre and terminus of both the Alaska Highway and the railway.

Northern Alaska

To the north of the Central Plateau is a range of bleak mountains known as the Brooks Range, rising to over 2400 metres above sea level. North of the mountains is a broad, low-lying plain bordering the Arctic Ocean. It is sometimes called the 'north slope' and much of it is tundra, grazed by musk ox and reindeer. There is a small Eskimo population engaged in the traditional occupations of hunting and fishing, and also in fur trapping and the herding of reindeer.

In the past there has been little economic activity, in spite of the known presence of coal and iron ore, but a remarkable new development may transform the whole way of life of the region. This has been the location of a vast oilfield deep down beneath the permafrost round Prudhoe Bay on the north coast. Large scale drilling is taking place

and results are so promising that work is to begin soon on laying a large-diameter pipeline to carry the oil to a warm water port in the south. Experiments are also taking place with icebreaker-tankers to find out whether the 'North-West Passage' can be used for marketing oil in the eastern United States, where it is mostly needed.

Exercises

Answer in note form :
1. Draw an annotated section from north to south across Alaska to show the physical features and land use.
2. Write about lumbering, fishing and mineral working in Alaska.
3. Compare the climates of Juneau and Fairbanks, and give reasons for the differences.
Essay :
   Why is the economic development of Alaska still in its early stages, and what possibilities exist for the future?